中国鼎
CHINA DING

2013

群星闪耀 盆景中国

中国盆景 年度之夜

将于 2013 年 9 月 29 日在广东省中山市古镇镇隆重推出

全体中国盆景人敬请期待!
主办:广东省中山市人民政府,中国盆景艺术家协会
承办:中山市古镇镇人民政府,中山市农业局,中山市林业局,
中山市海洋与渔业局,广东省盆景协会
协力支持单位:中山市南方绿博园有限公司

中国盆景赏石

2013-9

September 2013

中国林业出版社 China Forestry Publishing House

图书在版编目（CIP）数据

中国盆景赏石·2013. 9 / 中国盆景艺术家协会主编. -- 北京：中国林业出版社，2013.9

ISBN 978-7-5038-7187-0

Ⅰ. ①中… Ⅱ. ①中… Ⅲ. ①盆景－观赏园艺－中国－丛刊②观赏型－石－中国－丛刊 Ⅳ. ① S688.1-55 ② TS933-55

中国版本图书馆 CIP 数据核字（2013）第 212182 号

责任编辑：何增明 张华

出 版：中国林业出版社
E-mail:shula5@163.com
电话：（010）83286967

社 址：北京西城区德内大街刘海胡同 7 号
邮编：100009

发 行：中国林业出版社

印 刷：北京利丰雅高长城印刷有限公司

开 本：230mm×300mm

版 次：2013 年 9 月第 1 版

印 次：2013 年 9 月第 1 次

印 张：8

字 数：200 千字

定 价：48.00 元

主办、出品、编辑： 中国盆景艺术家协会

E-mail: penjingchina@yahoo.com.cn

Sponsor/Produce/Edit: China Penjing Artists Association

创办人、总出版人、总编辑、视觉总监、摄影：苏放

Founder，Publisher，Editor-in-Chief, Visual Director，Photographer : Su Fang

电子邮件：E-mail : 1440372565@qq.com

编辑报道热线 : 010-58693878（每周一至五：上午10 : 00-下午6 : 30）

News Report Hotline : 010-58693878 (10 : 00a.m to 6 : 30p.m, Monday to Friday)

传真Fax : 010-58693878

投稿邮箱Contribution E-mail : CPSR@foxmail.com

会员订阅及协会事务咨询热线 : 010-58690358（每周一至五：上午10 : 00-下午6 : 30）

Subscribe and Consulting Hotline : 010-58690358 (10 : 00a.m to 6 : 30p.m, Monday to Friday)

通信地址 : 北京市朝阳区建外SOHO16号楼1615室 邮编 : 100022

Address : JianWai SOHO Building 16 Room 1615, Beijing ChaoYang District, 100022 China

法律顾问：赵煜

Legal Counsel : Zhao Yu

制版印刷：北京利丰雅高长城印刷有限公司

读者凡发现本书有掉页、残页、装订有误等印刷质量问题，请直接邮寄到以下地址，印刷厂将负责退换：北京市通州区中关村科技园通州光机电一体化产业基地政府路2号 邮编101111

联系人王莉，电话：010-59011332。

铺地柏 *Juniperus procumbens* 高 78cm 获得“第十四届诺朗德斯杯‘比利时盆栽俱乐部会员最佳盆景’”
克里斯蒂安 · 沃斯藏品 供图：马克 · 诺朗德斯
Garden Juniper. Height: 78cm. Noelanders Trophy ‘Best Bonsai of a member of BAB XIV’.
Collector: Christian Vos, Photo Contributor：Marc Noelanders

中国盆景赏石 2013-9
CHINA PENJING & SCHOLAR'S ROCKS
September 2013

封面：山毛榉 *Fagus crenata* 高 110cm 获得"第 14 届诺朗德斯杯最佳盆景（落叶树种）" 路易斯·巴列霍藏品 供图：马克·诺朗德斯

Cover: Common Beech. Height:110cm. Noelanders Trophy 'Best deciduous bonsai XIV'. Collector: Luis Vallejo, Photo Contributor: Marc Noelanders

封四："古陶" 大化彩玉石 长 20cm 宽 16cm 高 24cm 李正银藏品 苏放摄影

Back Cover: "Antique Pottery". Macrofossil. Length: 20cm, Width: 16cm, Height: 24cm. Collector: Li Zhengyin, Photographer: Su Fang

卷首语 Preamble

海外景色 VIEW OVERSEAS

21 中欧山松 *Pinus mugo* 高 65cm 获得"第十四届诺朗德斯杯'比利时盆栽俱乐部会员最佳盆景'提名奖"彼得·吉艾伦藏品 供图：马克·诺朗德斯

Height: 65cm. Nomination for Noelanders Trophy 'Bonsai of a member of BAB XIV'. Collector: Peter Gielen, Photo Contributor: Marc Noelanders

22 中欧山松 *Pinus mugo* 高 60cm 获得"第十四届诺朗德斯杯'比利时盆栽俱乐部会员最佳盆景'提名奖"爱尔达·克里斯特尔斯藏品 供图：马克·诺朗德斯

Height: 60cm. Nomination for Noelanders Trophy 'Bonsai of a member of BAB XIV'. Collector: Alda Clijsters, Photo Contributor: Marc Noelanders

23 鸡爪枫 *Acer palmatum* 高 65cm 获得"第 14 届诺朗德斯杯'最佳落叶盆景'提名奖" 伍德·费舍尔藏品 供图：马克·诺朗德斯

Japanese Maple. Height: 65cm. Nomination for Noelanders Trophy 'Deciduous bonsai XIV'. Collector: Udo Fischer, Photo Contributor: Marc Noelanders

24 丝鱼川真柏 *Juniperus chinensis* 获得"第 14 届诺朗德斯杯'最佳小品盆景'提名奖" 马克·库珀和瑞塔·库珀藏品 供图：马克·诺朗德斯

Nomination for Noelanders Trophy 'Shohin XIV'. Collector: Mark and Ritta Cooper, Photo Contributor: Marc Noelanders

25 黑松 *Pinus thunbergii* 获得"第 14 届诺朗德斯杯'最佳小品盆景'提名奖"南希·威丽藏品 供图：马克·诺朗德斯

Japanese Black Pine. Nomination for Noelanders Trophy 'Shohin XIV'.
Collector: Nancy Vaele, Photo Contributor: Marc Noelanders

26 长白松 *Pinus sylvestris* 高 55cm 获得"杜塞尔多夫盆栽美术馆特别奖" 马里亚诺·爱莎藏品 供图：马克·诺朗德斯

Changbai Pine. Height: 55cm. Sonderpreis, Bonsai-museum Düsseldorf.
Collector: Mariano Aisa, Photo Contributor: Marc Noelanders

27 光叶榉 *Zelkova serrata* 获得"韩国小品盆栽协会获奖证书" 阿兰·沃特藏品 供图：马克·诺朗德斯

Japanese Zelkova. Certificate of Award, Korea Shohin Bonsai Association. Collector: Alain De Wachter, Photo Contributor: Marc Noelanders

28 长白松 *Pinus sylvestris* 高 65cm 获得"上海植物园获奖证书" 莫罗·斯丹姆伯格藏品 供图：马克·诺朗德斯

Changbai Pine. Height: 65cm. Certificate of Award, Shanghai Botanical Garden. Collector: Mauro Stemberger, Photo Contributor: Marc Noelanders

29 丝鱼川真柏 *Juniperus chinensis* 高 60cm 获得国际盆栽俱乐部优秀奖 莫罗·斯丹姆伯格藏品 供图：马克·诺朗德斯

Height: 60cm. Excellence Award, Bonsai Club International. Collector: Mauro Stemberger, Photo Contributor: Marc Noelanders

封面故事 Cover Story

30 马克·诺朗德斯的使命——专访比利时盆栽协会会长马克·诺朗德斯 访问者 摄影：苏放 受访者：马克·诺朗德斯 撰文：克里斯蒂安·沃斯

A Mission of Marc Noelanders—Conversation with Marc Noelanders the Chairman of Bonsai Association Belgium Interviewer & photographer: Su Fang Interviewee: Marc Noelanders Author: Christian Vos

海外现场 Spot International

34 紫杉变形记 设计和造型：马克诺朗德斯 文：马克德贝乌勒、艾丽泽曼、迈克尔埃克斯纳、约尔格·德林 翻译：约尔格·德林 摄影：威利和安琪·埃费内普尔

Metamorphosis of A Yew Design and Styling: Marc Noelanders Text: Marc de Beule, Elize Mann, Michael Exner, Jörg Derlien Translation: Jörg Derlien Photography: Willy and Antje Evenepoel

中国盆景赏石 2013-9
CHINA PENJING & SCHOLAR'S ROCKS
September 2013

国际盆景世界 Penjing International

盆景中国 Penjing China

会员之声 Member's Voice

专题 Subject

88 论盆景360°全景展示——罗汉松盆景"苍翠"的创作技术实现 文：关山

Talk About 360° Panoramic Display of Penjing—The Creation Technology of Yaccatree Penjing "Green" Author: Guan Shan

点评 Comments

90"向上" 对节白蜡 *Fraxinus hupehensis* 文：李奕祺

"Up-right" Author: Li Yiqi

养护与管理 Conservation and Management

92 盆景素材的培育（十四）——《盆景总论》（连载十六）文：【韩国】金世元

Penjing Materials Nurture — Pandect of Penjing (Serial XVI)

Author: [Korea] Kim Saewon

生活方式 Life Style

94 不是为了炫耀，而是为了人生——从瓦茨拉夫·诺瓦克的私人院落里的一块草坪谈起

摄影：苏放 撰文：CAT

For Life not Showing off —Talking From a Piece of Lawn in the Private Courtyard of Vaclav·Novak Photographer: Su Fang Author: CAT

古盆中国 Ancient Pot Appreciation

104 紫砂古盆铭器鉴赏 文：申洪良

Red Porcelain Ancient Pot Appreciation Author : Shen Hongliang

盆艺欣赏 Pot Art Appreciation

108 柯家花园仿古石盆系列欣赏

The Appreciation of the Ke Chengkun's Antique Pot Series

赏石中国 China Scholar's Rocks

109"天险" 九龙璧 长38cm 宽25cm 高45cm 魏积泉藏品

"Natural Barrier". Nine Dragon Jade. Length: 38cm, Width: 25cm, Height: 45cm. Collector: Wei Jiquan

110"貔貅" 大化彩玉石 长33cm 宽23cm 高13cm 李正银藏品 苏放摄影

"Pixiu". Macrofossil. Length:33cm, Width: 23cm, Height: 13cm. Collector: Li Zhengyin, Photographer: Su Fang

111"荡中寻宝" 广西天峨石 长15cm 宽11cm 高23cm 魏积泉藏品

"Treasure Hunt in the Swing". Tian'e Stone, Guangxi. Length: 15cm, Width: 11cm, Height: 23cm. Collector: Wei Jiquan

112 中国古今名石简谱（连载七） 文：文甡

Chinese Famous Rocks Notation(Serial Ⅶ) Author: Wen Shen

116 图纹石的质地之美解析 文：雷敬敷

The Analysis of the Beauty of the Texture of Figure Proluta

Author : Lei Jingfu

2013的中国时代

文: 苏放

2013-The Era of China

中国盆景赏石
CHINA PENJING
& SCHOLAR'S ROCKS
主编：中国盆景艺术家协会
Edited by China Penjing Artists Association
2011 1 Initial Issue

诺朗德斯杯无疑是欧洲盆景的一个亮点，如若想了解欧洲盆景，每年比利时的诺朗德斯杯是你应该去看的展览之一，我在诺朗德斯先生的盆景制作中看到了一个欧洲盆景人的内心历程，也听到了欧洲人与日本盆景的方向相异的另一种内心的声音，为了客观地向中国特别是亚洲盆景人展示欧洲盆景的近况和现实，我们首次为全球读者制作了来自比利时的诺朗德斯杯盆景展专辑，请读者们在把眼光投向欧洲盆景的时候品评一下诺朗德斯杯的展览，看看不同文化背景的诺朗德斯杯为当今的多元化世界又贡献了什么“不一样”的内涵。在阅读这本诺朗德斯杯专辑之前，请允许我向全球盆景人推荐这个展览。

中国盆景在 2013 年的 9 月将有 2 个值得瞩目的历史性事件，一个是在金坛开幕的首次由中国人来主办的世界盆景友好联盟大会，胡运骅先生担任了本届 WBFF 大会的主席，另一个是 9 月 29 日中国盆景艺术家协会（CPAA）在广东古镇推出的“2013 CHINA DING”中国鼎国家大展，这两个展览与之前扬州同样由中国人主办的 BCI 展览一道，向全世界盆景界宣告了 2013 中国时代的开始。

什么是 PENJING ？什么是 BONSAI ？很多外国人问过我。

中国是盆景的发源国，也是一个文化大国，但盆景的历史事实是：盆景为世界人民所知所爱是由于日本国家全民族的努力，并且经过日本人数百年坚韧不拔的努力，“BONSAI”已经成为以英语为母语的人所熟知的表述盆景含义的英文名词，这是日本盆景艺术家们的成功，也是日本文化的成功。在此，全世界的盆景人应该向日本盆栽表达敬意。

对很多外国人特别是欧美人来说，他们并不知道 PENJNG 与 BONSAI 表述的其实是一种东西，但却是两种文化，在中国的汉语中叫 PENJING，在中国之外的英语里叫 BONSAI。

2013 年来了，在中国，所有落地到中国的盆景包括外来盆景，如今都因为中国人独特的盆景审美哲学和选材标准，已经融汇到中国盆景中并成为中国盆景文化中的一部分。中国人对盆景美有自己非常独特的一套标准，这个标准永远与“YI JING”（意境）这个词有关。特别是中国“LING NAN PAI”（岭南派）的杂木类盆景的枝法更是当今盆景世界的一种全新的语言，这种语言只有在中国的广东省能大量地看到。

中国是一个遥远的带有东方文化色彩的幕帘之后的神秘国度，这个发明了盆景并拥有古老的自成体系的盆景文化和审美哲学的国家到底拥有什么样的盆景？虽然《中国盆景赏石》每本都会刊登一些中国盆景，但是图片和实物毕竟会有很大的区别，特别是很多连中国人自己都从未看到过的从未参加过任何展览的一些顶级的中国盆景现在到底什么样，2013 年所有的中国盆景展览，将揭开这一道最后的幕帘。

我想对全球读者说的话只有两句：1. 爱盆景，不看中国盆景，你不会知道盆景或 BONSAI 的文化根源和哲学本质来自何处；2. 不要相信图片和传说，中国的顶级盆景到底什么样，你将在 2013 年 9 月 25 日的中国江苏省的金坛市和 2013 年 9 月 29 日的中国广东省的中山市古镇镇的两个展览上找到全面的答案。

到金坛去！到古镇去！盆景的中国时代已经开始，中国盆景作为盆景起源国正在向全球发出一个邀请。

从扬州到金坛，再到古镇，从 BCI 到 WBFF 再到古镇的“2013 CHINA DING”中国鼎中国盆景国家大展，这是 3 个展览，中国盆景人用一个从北到南的地理轨迹，向全世界宣告了拥有辽阔地域和丰富多彩的盆景文化的中国时代的开始。

“YI JING”是理解中国盆景的一个核心关键词，但打开所有中英文字典，你无法找到一个英语为母语的人都知道的对应的英文词，我一直也苦于不知道如何表达这个词的准确翻译，有时只能大概地对外国朋友这样解释：像神枝和舍利干一样，“YI JING”是中国盆景特有的一个专有名词，不懂这个词你不会懂得什么是中国盆景。

是的，没来过中国、没看过中国盆景的人，你不会知道“YI JING”一词的含义。

你不会知道盆景的根源和本质来自何处。

你无法知道中国盆景的背后的传统和文化在表达什么。

你不会知道在日本盆景诞生之前中国人发明了盆景是因为什么。

你不会知道盆景在中国人的生活方式、美学和哲学中与大自然的那种“天人合一”的关系是什么意思？

你更不会知道为什么中国人不把盆景看成是园艺而是艺术。

中国人一千多年前就用盆景、盆景盆、盆景几架的三位一体发现了盆景的仪式感和其背后的文化内涵，这是中国人对盆景的本质的一种发掘和表达。要了解这种表达和文化，你应该到盆景的发源国——中国来看一下，亲眼看看“YI JING”这个词的含义到底是什么，中国盆景与你已经习惯看到的其他国家的盆景到底有什么不同？

所以，2013 年 9 月，到中国来吧，中国的展览将告诉你这一切的答案。

特别是：什么叫“YI JING”？

由“YI JING”这个词开始，你将开始真正地了解中国盆景和其背后的文化。

最后，再建议大家，用掌声祝贺比利时的诺朗德斯杯。

2013 - The Era of China

Author & Photographer: Su Fang

No doubt, the Noelanders Trophy is a highlight of European Penjing. It is an exhibition you have to visit if you want to know European Penjing. I see the inner journey of one European Penjing artist through Mister Noelanders' Penjing creation. I also heard European's inner voice different from Japanese Bonsai. In the aim of introducing the recent situation of European Penjing objectively to Chinese especially Asian Penjing people, we first publish the album about the Noelanders Trophy from Belgium. Readers please take a look at the European Penjing and see how the Noelanders Trophy with different background contributes to the world. Please let me recommend this exhibition.

There will be two great events of China Penjing in September, 2013. One is the World Bonsai Friendship Federation sponsored by the Chinese people for the first time, with Hu Yunhua as the president. The other is China Ding 2013 China National Penjing Exhibition launched by China Penjing Artists Association. These two exhibitions together with BCI exhibition sponsored by Chinese people too declared to the whole world that the beginning of China Era.

Many foreigners ask me: "What is Penjing? What is Bonsai?"

China is the cradleland of Penjing and also a great cultural nation. But the fact about Penjing history is that Penjing can be known to the whole world owe much to the efforts of the whole nation of Japan and now "Bonsai" becomes an English word being familiar with native English speakers owing to Japanese persistent efforts for hundreds of years. This is a success of Japanese Bonsai artists and also a success of Japanese culture. Here, all Penjing people all over the world should pay our respects to Japanese Bonsai.

For many foreigners especially for Europeans and Americans, they have no idea that Penjing and Bonsai represent one object but two different cultures. People in China call it Penjing and outside China people call it Bonsai in English.

We are in the year of 2013. All Penjing either from local or foreign countries that settle in China all have been affected by the influence of Chinese's unique aesthetic philosophy and selection standard, now they blend into China Penjing and become a part of it. Chinese have a unique set of standard towards the beauty of Penjing and this standard has always been related to the Chinese word "YI JING". Especially the technique of miscellaneous Penjing of Chinese "Lingnan Pai" which is a whole new language of Penjing world and can be found only in Guangdong Province, China.

China is a mysterious nation with oriental culture. In this nation which invents Penjing and have a system of Penjing culture and aesthetic philosophy of its own, what kind of Penjing you can find in nation like this? Although "China Penjing & Scholar's Rocks" will publish pictures about China Penjing on each issue, there is still a difference between pictures and real object. Some top Penjing which had never been into any exhibition even Chinese people do not know them.

What I want to say to global readers are : 1. Although you love Penjing, you do not know China Penjing, you will find it difficult to know the origin of culture and nature of philosophy of Penjing and Bonsai. 2. Do not believe pictures and rumors. What are China's top Penjing like? You will find out in the two exhibition held in Jintan City, Jiangsu Province, China on September 25th, 2013 and Guzhen Town, Zhongshan City, Guangdong Province, China on September 25th, 2013.

Go to Jintan! Go to Guzhen! The era of China Penjing is beginning. China as the cradleland of Penjing is sending an invitation to the whole world.

From Yangzhou, Jintan to Guzhen. From BCI, WBFF to "2013 China Ding". These three exhibitions from north to south have declared the beginning of an era of China the nation with vast territory and colorful Penjing culture.

"YI JING" is a keyword to understand China Penjing. When you looked it up in the dictionary, you will not find any corresponding English words that English speakers know. I have been struggling with the translation of this Chinese word for a long time, sometimes I can just roughly explain to others: like God branch and Sarira, "YI JING" is a special term of Penjing culture. You cannot understand China Penjing until you learn this word.

Yes, you will not understand the meaning of "YI JING" if you have never been to China and seen China Penjing.

You will not know where the root and the essence of China Penjing coming from.

You will not know what are the tradition and culture hidden behind China Penjing trying to express.

You will not know the reason why Chines invented Penjing before the birth of Japanese Bonsai.

You will not know the meaning of the relation between Chinese life style, aesthetics, philosophy and nature's concept "TianRenHeYi".

And still you would not know why Chinese regard Penjing as art instead of gardening.

A thousand years ago, Chinese people had already used Penjing, Penjing pot and Penjing shelves to express a sense of ritual and cultural connotation. For Chinese, this is a way of discovering and expressing the nature of Penjing. In order to understand this kind of culture and this way of expression, you better come to China and learn the word "YI JING" by yourself. Then you can figure out the difference between China Penjing and Other countries'.

So, come to China in September, 2013. You will get all the answer from China's exhibition.

Especially: What is "YI JING"?

From the word "YI JING", you will begin to truly understand China Penjing and the culture behind it.

In the end, I suggest we congratulate on the success of the Noelanders Trophy with applause.

Go to Jintan! Go to Guzhen!
The era of China Penjing is beginning.
China as the cradleland of Penjing is sending an invitation to the whole world.

2013 年から、中国のすべての盆景（海外からの盆景も含め）は中国人の独特な盆景審美哲学と木材の選択標準によって中国盆景と融し、もう中国盆景文化の一部分となった。中国人は盆景の美意識について自分の非常に独特な標準が有り、そして、この標準がずっと「意境」という言葉に関する。特に、中国嶺南派の雑木盆景の技法は今の盆景世界の真新しい言語の一種で、中国の広東省だけでよく見られる。

中国は東方文化色彩を持って遠くの神秘な国である。この盆景を発明し、盆景文化と審美哲学を持つ国は、いったいどのような盆景があるのか？『中国盆景賞石』で毎期数ページの中国盆景写真が載せられるけど、何といっても写真と本物は大きな区別が有るので、本物を見た後もっと了解すると思う。今、展覧会に参加したことがなく、更に中国人でも見なかった最高盆景の真面目が今年の展覧会を通して呈するのであろう。

世界の読者に言いたいことは二句だけ、1。もし盆景が好きだけど、中国盆景を見たことがないなら、盆景と盆栽の文化根源と哲学本質に分かるわけがない；2。写真と伝説を信じられない。2013 年 9 月 25 日に中国江蘇省金壇市と 2013 年 9 月 29 日に中国広東省中山市古鎮町の展覧会からすべての答えが探される。

金壇へ行く！古鎮へ行く！盆景の中国時代がもう始まり、中国盆景は盆景の発源国として全世界に招待している。

揚州から金壇へ、また古鎮へ行く。BCI 大会からＷＢＦＦ大会、また、"2013 CHINA DING" 中国鼎国家大展までの三つの展覧会が、北から南への地理軌跡で、中国人は広い地域と豊かな盆景文化を持つ中国時代の始まりと世界に伝える。

「意境」は中国盆景を理解させられるキーワードである。しかし、中英辞典の中で外国人に理解させる英語がなく、私もいかにこの言葉を正確に翻訳すると苦しんでいる。時には、外国の友達に「ジンとシャリのようなこと、意境は中国盆景の専門用語だ。」とおおざっぱに説明する。

中国へ行ったことがなく、中国盆景を見たことがない人にとって、意境という言葉の意味がなかなか理解できない。

盆景の根源と本質はどこから生まれるのが分からない。

中国盆景の裏側の背景と文化は何のことを現したいとも分からない。

日本盆栽の誕生の前に中国人が盆景を発明した原因も分からない。

盆景は中国人の生活方法・美学・哲学の中で自然との「天人合一」の関係が知らない。

どうして中国人は盆景が園芸にしなく、芸術とすることも知らない。

千年前に中国人は盆景・鉢・棚の三位一体を通して、盆景の儀式感とその裏側の内包を感じた。それは中国人が盆景に対して本質の発掘と表現することである。もし、このような表現と文化を理解したいなら、盆景の発源国—中国へいらっしゃると勧めたい。ここで、「意境」という言葉の意味とか、他国の盆景との違いとか、自分で探求される。

だから、2013 年 9 月の際、中国に来るよ。中国の展覧会がすべて教えてくれる。

特に、意境はどのようなことか？

意境という言葉から、中国盆景とその裏側の文化を確かに了解し始めようと思う。

末筆ながら、拍手でベルギーのノーランダスカップを祝うと提案する。

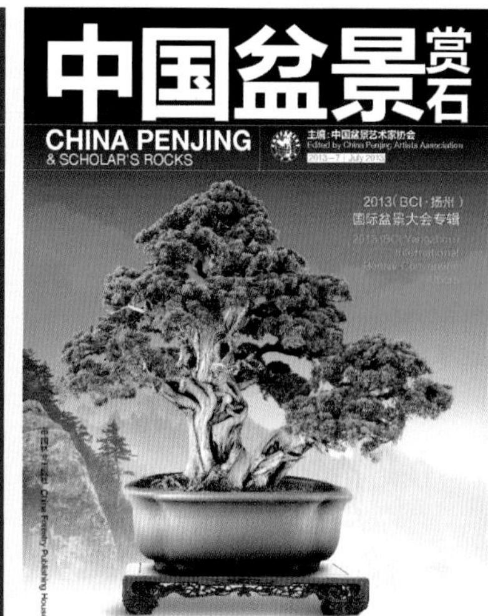

2013の中国時代

文：蘇放

ノーランダスカップ（Noelanders Trophy）はヨーロッパ盆景の目玉事情の一つといっても言い過ぎではない。ヨーロッパ盆景を了解したいと、年一回のノーランダスカップが勧められる展覧会の一つである。ノーランダス氏のデモンストレーションから一つのヨーロッパ盆景人の心の旅が見えたり、日本盆栽と違う心の声が聞けたりした。中国及びアジアの盆景人にヨーロッパ盆景の近況と現実を客観的に紹介するために、今期ベルギーのノーランダスカップ事情を載せて、読者がヨーロッパ盆景に注目を集める時にノーランダスカップの作品にも鑑賞させ、異なる文化背景のノーランダスカップが今多元化の世界に何か異なる内包を貢献したことを見せられてほしい。今期のノーランダスカップ作品集を御覧前に、まず、今年中国の展覧会をご紹介。

2013年9月、中国盆景は一見に値する歴史的な事情が二つある。一つは金壇にて開幕の世界盆栽友好連盟大会、これは初めて中国人によって開催され、胡運驊氏が今回のWBFFの会長になる。一つは9月29日に中国盆景芸術家協会（CPAA）が広東古鎮にて主催する"2013CHINA DING"中国鼎国家大展である。この二つの展覧会は、数月前に同じ中国において主催されたBCI大会と一緒に、全世界の盆景界に2013は中国時代の始まりと告げる。

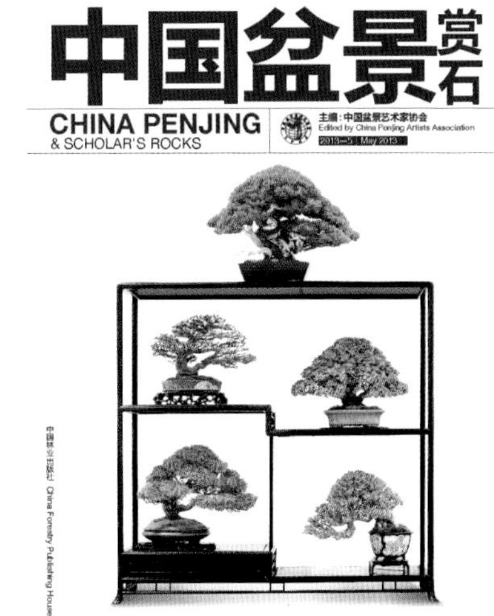

盆景は何？盆栽は何？と聞いた外国人が多い。

中国が盆景の発源国であり、その共に文化大国であるが、盆景の歴史事実は：盆景が世界の人々に知られて好きになるのは、日本国全民族の努力と、日本人がそのために数百年間のたゆまず頑張ることである。「BONSAI」は英語の母語の人たちによく知られる言葉で盆景と表示する。これは日本盆栽芸術家の成功、それとも日本文化の成功である。ここで、全世界の盆景人が日本盆栽に敬意を表すと思う。

多くの外国人、特に欧米人にとって、盆景と盆栽が同じ種類の物を表示するが、実は二つ種類の文化だとは知らない。中国の漢語で「PENJING」と読んで、中国以外の英語で「BONSAI」と読む。

鱼鳞云杉 *Picea jezoensis* 高 80cm 获得“第 14 届诺朗德斯杯”
路易斯 · 巴利尼诺藏品 供图：马克 · 诺朗德斯
Little-sperm Yeddo Spruce. Height: 80cm. Noelanders Trophy XIV.
Collector: Luis Balino, Photo Contributor：Marc Noelanders

赤松 *Pinus densiflora* 获得“第 14 届诺朗德斯杯‘最佳小品盆景’”
马克 · 库珀和瑞塔 · 库珀藏品 供图：马克 · 诺朗德斯
Japanese Red Pine. Noelanders Trophy ‘Best Shohin XIV’.
Collector: Mark and Ritta Cooper, Photo Contributor：Marc Noelanders

五叶松 *Pinus pentaphylla* 高 80cm 获得“第 14 届诺朗德斯杯提名奖”
路易斯 · 巴列霍藏品 供图：马克 · 诺朗德斯
Height: 80cm. Nomination for Noelanders Trophy XIV.
Collector: Luis Vallejo, Photo Contributor：Marc Noelanders

油橄榄 *Olea europea sylvestri* 高 74cm 获得“第 14 届诺朗德斯杯提名奖”
加西亚 · 费尔南德斯 · 艾拉斯莫藏品 供图：马克 · 诺朗德斯
Olive. Height: 74cm. Nomination for Noelanders Trophy XIV.
Collector: Garcia Fernandez Erasmo, Photo Contributor：Marc Noelanders

五叶松 *Pinus pentaphylla* 高 72cm 获得“第 14 届诺朗德斯杯提名奖”
佩德罗 · 佩斯藏品 供图：马克 · 诺朗德斯
Height: 72cm. Nomination for Noelanders Trophy XIV.
Collector: Pedro Paes Mamede, Photo Contributor：Marc Noelanders

石榴 *Punica granatum* 高 80cm 获得“第 14 届诺朗德斯杯提名奖”
弗雷德里克 · 夏奈尔藏品 供图：马克 · 诺朗德斯
Pomegranate. Height: 80cm. Nomination for Noelanders Trophy XIV.
Collector: Frederic Chenal, Photo Contributor: Marc Noelanders

长白松 *Pinus sylvestris* 高 67cm 获得“第 14 届诺朗德斯杯提名奖”
马里亚诺 · 爱莎藏品 供图：马克 · 诺朗德斯
Changbai Pine. Height: 67cm. Nomination for Noelanders Trophy XIV.
Collector: Mariano Aisa, Photo Contributor：Marc Noelanders

中欧山松 *Pinus mugo* 高 65cm 获得“第十四届诺朗德斯杯‘比利时盆栽俱乐部会员最佳盆景’提名奖”
彼得 · 吉伦藏品 供图：马克 · 诺朗德斯
Height: 65cm. Nomination for Noelanders Trophy ‘Bonsai of a member of BAB XIV’.
Collector: Peter Gielen, Photo Contributor：Marc Noelanders

中欧山松 *Pinus mugo* 高 60cm 获得“第十四届诺朗德斯杯‘比利时盆栽俱乐部会员最佳盆景’提名奖”
爱尔达 · 克里斯特尔斯藏品 供图：马克 · 诺朗德斯
Height: 60cm. Nomination for Noelanders Trophy ‘Bonsai of a member of BAB XIV’.
Collector: Alda Clijsters, Photo Contributor：Marc Noelanders

鸡爪枫 *Acer palmatum* 高 65cm 获得“第 14 届诺朗德斯杯‘最佳落叶盆景’提名奖”
伍德 · 费舍尔藏品 供图：马克 · 诺朗德斯
Japanese Maple. Height: 65cm. Nomination for Noelanders Trophy ‘Deciduous Bonsai XIV’.
Collector: Udo Fischer, Photo Contributor：Marc Noelanders

丝鱼川真柏*Juniperus chinensis* 获得“第14届诺朗德斯杯‘最佳小品盆景’提名奖”
马克 · 库珀和瑞塔 · 库珀藏品 供图：马克 · 诺朗德斯
Nomination for Noelanders Trophy ‘Shohin XIV’.
Collector: Mark and Ritta Cooper, Photo Contributor：Marc Noelanders

黑松 *Pinus thunbergii* 获得“第 14 届诺朗德斯杯‘最佳小品盆景’提名奖”
南希 · 威丽藏品 供图：马克 · 诺朗德斯
Japanese Black Pine. Nomination for Noelanders Trophy ‘Shohin XIV’.
Collector: Nancy Vaele, Photo Contributor：Marc Noelanders

长白松 *Pinus sylvestris* 高 55cm 获得“杜塞尔多夫盆栽美术馆特别奖”
马里亚诺 · 爱莎藏品 供图：马克 · 诺朗德斯
Changbai Pine. Height: 55cm. Sonderpreis, Bonsai-museum Düsseldorf.
Collector: Mariano Aisa, Photo Contributor：Marc Noelanders

光叶榉 *Zelkova serrata* 获得“韩国小品盆栽协会获奖证书”
阿兰 · 沃特藏品 供图：马克 · 诺朗德斯
Japanese Zelkova. Certificate of Award, Korea Shohin Bonsai Association.
Collector: Alain De Wachter, Photo Contributor：Marc Noelanders

长白松 *Pinus sylvestris* 高 65cm 获得“上海植物园获奖证书”
莫罗·斯丹姆伯格藏品 供图：马克·诺朗德斯
Changbai Pine. Height: 65cm. Certificate of Award, Shanghai Botanical Garden.
Collector: Mauro Stemberger, Photo Contributor：Marc Noelanders

丝鱼川真柏 *Juniperus chinensis* 高 60cm 获得国际盆栽俱乐部优秀奖
莫罗 · 斯丹姆伯格藏品 供图：马克 · 诺朗德斯
Height; 60cm. Excellence Award, Bonsai Club International.
Collector: Mauro Stemberger, Photo Contributor：Marc Noelanders

A Mission of Marc Noelanders

马克·诺朗德斯的使命

——专访比利时盆景协会会长马克·诺朗德斯

— Conversation with Marc Noelanders the Chairman of Bonsai Association Belgium

访问者 摄影: 苏放 受访者: 马克·诺朗德斯 撰文: 克里斯蒂安·沃斯
Interviewer & photographer: Su Fang Interviewee: Marc Noelanders Author: Christian Vos

问：作为欧洲盆景协会的副会长，您能简单地谈一下欧洲盆景的历史吗？哪3个国家能代表当代欧洲盆景的最高水平？

答：欧洲盆景的历史最早可以追溯到1862年，在此时期日本参加了在英国伦敦举办的世界博览会，以及之后1879年奥地利维也纳世界博览会，1876年和1900年在法国巴黎举办的两届世界博览会。这大概就是欧洲人最早与盆景结缘的阶段。但是，欧洲人真正开始了解盆景是在1945~1950年。

一些欧洲盆景大师不远万里去到日本学习盆景艺术。我就在日本学习了3年盆栽艺术，然后发展出了自己的个人风格。

在欧洲，盆景文化比较繁荣的国家有英国、德国和意大利。因为这些国家都是多山国家，人们很容易找到好的坯材来制作盆景。

在最初阶段，盆景都是从国外引入的，大多来自日本。当然，也有一些盆景艺术家会使用本国当地的树种。他们中的大多数都是从中国和日本的画作中汲取盆景制作灵感。

我们可以说，现今西班牙的盆景艺术水平也很高。事实上，在欧洲许多国家都可以找到技术精湛的盆景艺术家。

问：请简单介绍下诺朗德斯杯的现状？您认为它对欧洲盆栽的意义和价值是什么？

答：2014年1月18~19日两天将举办第15届“诺朗德斯杯”。在全欧洲范围来讲，它可以被认为是最重要的盆景展览之一（或许是最重要的）。

每届“诺朗德斯杯”的参观者数量大约有2000人。展览上的展品是选自全欧洲最顶级的120盆盆景。展会过程中我们将邀请4~6名盆景制作表演者向大家现场展示技艺。表演者来自德国、法国、西班牙、意大利、捷克共和国、英国、美国、印度尼西亚等国家。此外，在展览上还会有30名左右的贸易商，盆景这项爱好所需要的所有东西在这里都可以买到：盆景工具、盆景盆等。

我们可以骄傲地说，无论对于盆景专业人士还是业余人士，能够入选“诺朗德斯杯”就是一项荣誉。

问：您创办“诺朗德斯杯”的目标是什么？到达没有？

答：“诺朗德斯杯”每年一届，由比利时盆景协会主办。我们协会的使命就是提高比利时盆景的档次和文化，并面向世界展示比利时盆景。

我们现在还不敢说已经达成了这个目标。

问：从这几届“诺朗德斯杯”展览上看，当代欧洲盆景的发展轨迹是什么样的？未来会朝着什么方向发展？

答：经过了14届“诺朗德斯杯”，我们能看到盆景品质的不断提高。10年之前，高水平的盆景作品并不多见。现在，多亏了出现在欧洲的盆景工作坊和盆景展览对盆景文化的推动，欧洲盆景艺术家的技艺已经达到了一个新的高度。

当然，也离不开现代兴起的多种通信方式的帮助，比如因特网和脸谱网让人们的联系更加紧密。这就让年轻的盆景业余爱好者可以通过网络欣赏到更多使人惊艳的盆景作品并且从中得到灵感受到启发。

我们希望在这个充满压力的世界里，更多的人通过制作盆景来寻找心中的那份宁静。

问：在亚洲，大部分盆景收藏者都是有钱人或企业家，欧洲的盆景收藏群体是怎么样的一群人？

答：在欧洲，一方面我们有以盆景产业谋生的专业盆景艺术家。他们中有的人有盆景园和盆景商店，可以卖树、卖工具等，还会举办工作坊。另一方面，我们还有把盆景完全当做业余爱好的人，只是怀着单纯的热情，做的盆景却相当不错。

欧洲几乎没有非常有钱的收藏者，因为这样的人要雇很多人打理他的收藏品。以美国费城外著名的肯尼特广场盆景收藏展为例，它的盆景的拥有者是道格拉斯·保罗先生，像这种规模的展览，甚至规模比它小，而且藏品完全属于个人的情况，在我看来在欧洲根本不存在。但是，一些欧洲的盆景艺术家如果有珍贵的藏品的话，大部分时间也会向公众开放参观。

比如说：

马德里阿尔科文达斯盆景展（西班牙）、米兰克雷斯皮盆景博物馆（意大利）、伯明翰国家盆景收藏展（英国）。

问：您觉得对欧洲人来说盆景是什么：园艺？艺术品？您本人是怎么看的？

答：从一个特定的层面来说，盆景可以被说成是一门艺术就像不是所有的画作都是艺术品一样。但是，更重要的一点是，盆景拉近了人与自然的联系，使人们可以尽情享受工作或是业余爱好所带来的快乐。

问：通过我们寄给您的这本《中国盆景赏石》——中国盆景在您的印象中是什么样的？您觉得中国盆景与日本盆栽和欧洲盆景的区别是什么？

答：中国盆景与日本盆栽和欧洲盆景在风格上肯定有区别。但是每种风格都是平等的，没有孰好孰坏。也许，提起中国盆景想到的更多的是山水盆景。在欧洲，我们不会在盆景中布置小配件比如鹿、渔翁等。

不得不提到一点，最开始欧洲盆景艺术家的灵感来源于中国和日本的盆景并受其风格的影响。逐渐地他们开始创造出个人风格，也由于这个原因，他们的盆景选材都是欧洲当地的树种。

问：您怎么看未来10年整个世界盆景的发展？

答：这是一个非常难回答的问题……谁能知道10年后会发生什么呢？看上去日本的盆栽界很少有年轻人参与进来，当然中国盆景和日本盆栽的发展远远先于欧洲，两国都有很多树龄很老的盆景和技艺很高的盆景大师。我们希望欧洲年轻的盆景艺术家和业余爱好者都能参与进来使盆景事业更加繁荣。

问：您为什么特别注重盆景制作中的戏剧性，请您结合您的盆景制作谈谈您的这种观念好么？

答：可以说，我比较喜欢创造新事物、新风格，不论是小盆景还是超级盆景，我都想创造出独特的风格。任何树种对我来说都是一样的，无论是松柏、落叶树还是热带树。当然，技术和对盆景艺术的投入也是十分重要的。当我看到一棵树的时候，我的脑海里马上就会出现制作完成后整个盆景造型的画面。（注：马克·诺朗德斯是第一个被日本盆栽界承认为大师的欧洲盆景艺术家）

问：您有什么想对中国盆景人和我们这本书说的话吗？

答：中国盆景有自己的灵魂，独特的风格，长久的中国传统使得中国盆景变得这么的纯粹、这么的艺术。全世界不同文化、不同语言、不同信仰、不同生活方式的人们因为盆景又或叫盆栽而聚到一起，创造更好的世界。

《中国盆景赏石》这本书的内容十分有趣，通过阅读它可以了解好多东西。

The Noelanders Trophy is a yearly held exhibition organized by the Bonsai Association Belgium with Marc Noelanders as President. The mission of our association is to raise the class and knowledge of bonsai in Belgium and beyond the borders of the country.

Q: As the vice-chairman of European Bonsai Association, could you please briefly describe the history of European bonsai. Which are the three countries that can represent bonsai's highest level in Europe?

A: We can say that the origin of the European bonsai goes back to 1862 when Japan participated at the London (U.K.) World Exhibition and afterwards at the World Exhibition in 1879 in Vienna (Austria) and the two World Exhibitions held in Paris (France) in 1876 and 1900.

This is probably the period that Europeans came in contact with bonsai. But it is only since 1945 ~1950 that bonsai started to be known.

Some bonsai Masters went to Japan to be taught the art of bonsai. I went to Japan for 3 years and afterwards I developed my personal style.

The main countries in Europe for bonsai are U.K., Germany and Italy. Seen these countries have many mountains, people doing bonsai could easily find good raw material (yamadori).

In the beginning some bonsai were imported, mainly from Japan, other bonsai artists worked on local trees. They found their inspiration from drawings and paintings from China and Japan.

We can say that now Spain also has reached a high level of bonsai. But in fact some good bonsai artist can be found in most of the European countries.

Q: Please give us a brief introduction of the present situation about Noelanders Trophy. What do you think of Noelanders Trophy's meaning and value to European bonsai?

A: The Noelanders Trophy, which will have its 15th edition on January 18 and 19, 2014, can be considered as one of the most important bonsai shows in Europe (probably the most important one).

There are about 2000 visitors of the Noelanders Trophy.There are about 120 selected top bonsai from all over Europe on the show. We invite 4 to 6 demonstrators showing their skills. These demonstrators come from Germany, France, Spain, Italy, Czech Republic, U.K., U.S.A., Indonesia, and so on.

Moreover, on the show there are about 30 traders selling everything one needs for his hobby: pre-bonsai, tools, pots, a.s.o.

We can proudly say that it is an honor for bonsai professionals and amateurs to have their bonsai to be selected to participate at the Noelanders Trophy.

A Mission
of Marc Noelanders

Q: What is the purpose of founding Noelanders Trophy? Do you think you have already achieved the goal?

A: The Noelanders Trophy is a yearly held exhibition organized by the Bonsai Association Belgium with Marc Noelanders as President. The mission of our association is to raise the class and knowledge of bonsai in Belgium and beyond the borders of the country.

We do not consider that we achieved our goals: there is still a lot to do and we go for it.

Q: Based on several Noelanders Trophy exhibitions, what do you think the development path of contemporary European Bonsai? What is the direction in its future development?

A: Over the 14 years of the Noelanders Trophy we can see a development of the quality of bonsai.

10 years ago high standard bonsai were only present in smaller quantities. Nowadays, thanks to many workshops and exhibitions which take place in Europe, we can say that European bonsai artists have reached a high quality level.

Of course modern communication possibilities like internet and Facebook bring people in closer contact. This gives younger bonsai amateurs the possibility to admire and be inspired by magnificent bonsai shown on the web.

We hope more people living in this stressful world will enjoy and relax by practicing bonsai.

Q: In Asia, most of Penjing collectors are rich man or entrepreneurs, what are European bonsai collectors like, could you please describe them?

A: In Europe we have professional bonsai artists who do this job for a living. They sometimes have bonsai nursery and a shop where they sell trees, tools, a.s.o. and they are doing workshops.

On the other hand we have bonsai amateurs who are doing bonsai as a hobby, sometimes with passion and very good results.

Rich collectors nearly don't exist in Europe because such a person would need people to maintain this collection.

Take the famous Kennett Square Bonsai Collection in the U.S., just outside of Philadelphia, own by Mister Douglas Paul as an example. Such thing, or even smaller, own by private collectors, do in my opinion not exist in Europe. But some bonsai artists have good collections and are most of the time open to public for visiting.

Some examples:

Alcobendas Bonsai in Madrid (Spain),

Crespi Bonsai Museum in Milan (Italy),

The National Bonsai Collection Birmingham (UK)

Q: What does bonsai mean to European: Gardening? Art? Personally, what do you think of it?

A: Starting from a certain level, bonsai is art just like not all paintings are pieces of art.

But most important is that bonsai brings people in contact with nature and that they can enjoy their job or hobby.

Q: What do you think of China Penjing based on the book we sent to you? What is the biggest difference between China Penjing, Japanese Bonsai and European Bonsai?

A: Of course there is a difference in style between China Penjing and Japanese or European bonsai. But both styles are equivalent. Maybe Chinese Penjing represents more landscaping. In Europe we do not add some objects in the landscape as deer, fishermen, a.s.o.

It must be noted that in the beginning European bonsai artists were inspired by Chinese and Japanese bonsai styles. Little by little they created their own style also due to the fact that they worked on local species.

Q: What do you think of the development of the bonsai world in the next 10 years(Direction, artistical value, cultural values, market values, the difference between countries and difference between Asia and Europe? Outlook and expectation of technical progress?)

A: Difficult question … Who can look 10 years in advance?

It seems that not so much young people are involved in bonsai in Japan, but of course China and Japan started long before Europe with bonsai so both have very old bonsai and there are some famous bonsai Masters. We hope that in Europe young bonsai amateurs and artists will emphasize bonsai.

Q: Why do you put special emphasis on drama in bonsai creation, would you like to talk about this concept about drama based on your bonsai creation experience?

A: I like to create new things, new styles from small to extreme huge bonsai. I am able to create a bonsai out of any kind of tree, pines, deciduous trees, tropical trees, … Of course my skill and artistic input are important too.

When I look at a tree, I immediately have an image of the final work.(By the way, Marc was the first European bonsai artist who has been accepted as a Sensei in Japan)

Q: Is there something you want to say to China Penjing people and our book?

A: Chinese Penjing have their own style, their own soul, it is pure, it is art due to long Chinese tradition. One very important thing is that Penjing/Bonsai are bringing people from over the world with their different culture, different language, different religion, different way of living together to make a better world.

We want to congratulate China Penjing and scholar 's rocks with your magnificent book full of interesting articles. It is fantastic to learn so much about Penjing trough your book.

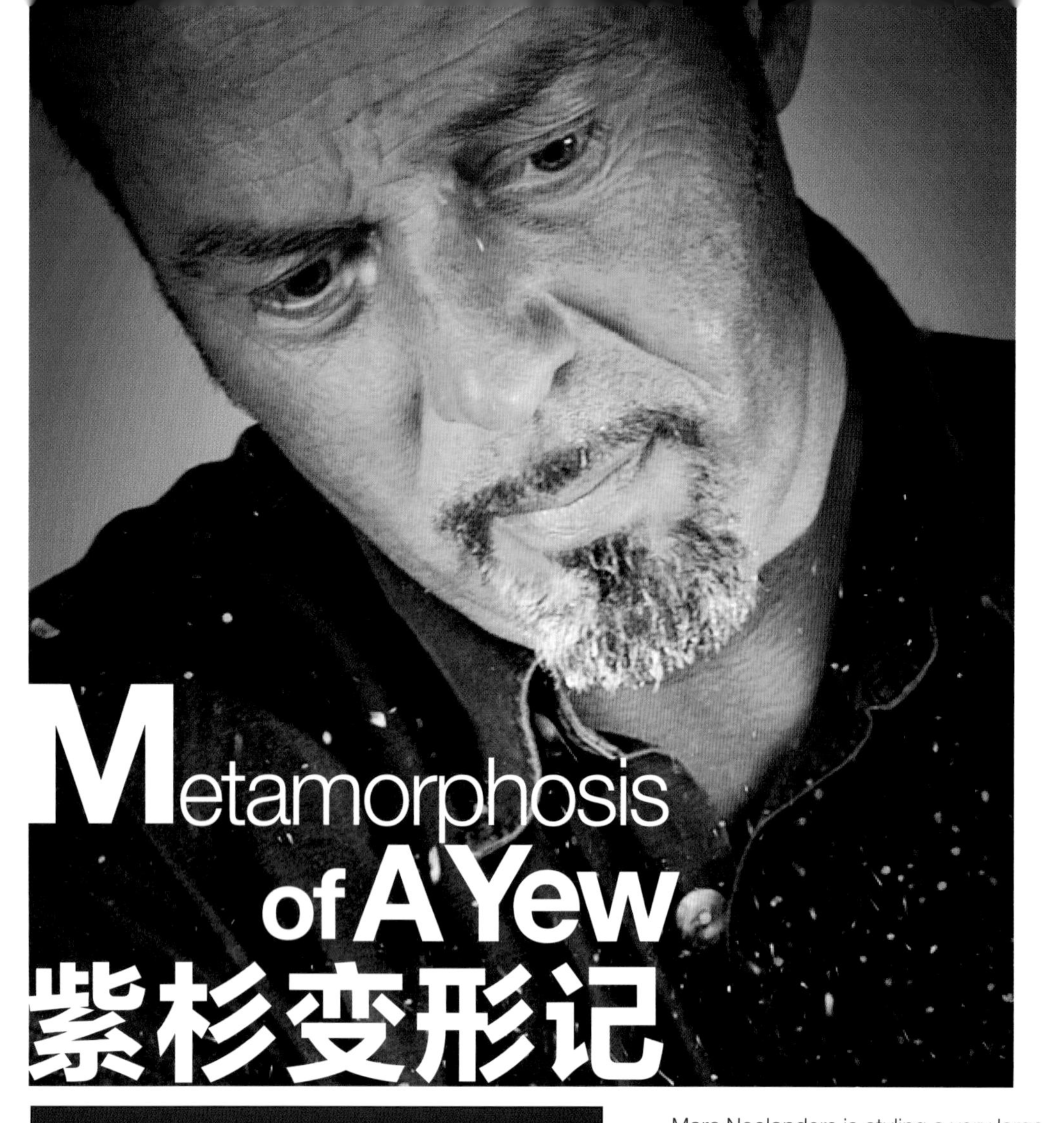

Metamorphosis of A Yew
紫杉变形记

马克·诺朗德斯打算给一棵100多年树龄的大型紫杉改天换地。

很多年以前，比利时盆景大师马克·诺朗德斯就已经从事盆景创作了。他曾在日本学习盆栽技艺，是木村正彦的学生。对他来说，提高专业技能至关重要。如今，他游遍世界各地、成立工作室、进行示范表演，并在其所属院校从事教学工作。他还是欧洲盆景协会(EBA)的副会长，也是欧洲最大盆景秀之一——"诺朗德斯杯"的创始人。

在本文中，他将展示他高超的技艺，将一棵年老的景观树紫杉(欧洲红豆杉)改造成一盆极好的盆景。

设计和造型：马克·诺朗德斯
文：马克·德·贝乌勒、艾丽泽·曼、迈克尔·埃克斯纳、约尔格·德林
翻译：约尔格·德林
摄影：威利和安琪·埃费内普尔
Design and Styling: Marc Noelanders
Text: Marc de Beule, Elize Mann, Michael Exner, Jorg Derlien
Translation: Jorg Derlien
Photography: Willy and Antje Evenepoel

Marc Noelanders is styling a very large and more than 100 years old yew.

The Belgium bonsai-master Marc Noelanders is doing bonsai since many years. He learned bonsai in Japan and was a student of Masahiko Kimura. For him it is very important to enhance his professional skills. Today he is traveling all over the world and is doing workshops, demonstrations and he teaches in his academies. Also he is the vice president of the Europe Bonsai Association (EBA) and founder of one of the biggest bonsai-shows in Europe the "Noelanders Trophy".

In this article he will proof his skills and transform an old garden yew (Taxus baccata) into a great bonsai.

The challenge

Traditionally many bonsai masterpieces are formed out of yamadori. The main work is done by the nature itself and the bonsai master only has to refine this impressive material.

Another way to do bonsai, is to use material from normal nurseries. But it's a long way from a seed to a good bonsai. Sometimes you can find raw material in gardens. But this plants usually neither have the impressive character of a yamadori, nor the structure of a nursery plant.

However, Marc Noelanders is going to show that he can make bonsai out of a 5 meter high and 50cm thick garden tree. The first challenge was to dig the really big tree and to pot it into a pot. The yew recovered very quickly and only 2 years later he was able to repot the tree into a bonsai pot. The old thick roots died and many fine roots grew very quickly. The trunk was almost 150cm high and had no tapering. The alive branches grew above a height of 60~100cm. The question now was, how it would be possible to create a bonsai out of a material like this.

图1(左图)原材料，欧洲紫衫(欧洲红豆杉)，高约150cm，树干直径50cm。造型前的正面
The raw material, European yew (*Taxus baccata*), height is about 150 cm, trunk diameter is 50 cm. The front side before styling

图 2 造型前的右侧
Right side before styling

图 3 造型前的左侧
Left side before styling

图 4 造型前的后方
Rear side before styling

挑战

传统上讲，许多盆景杰作都出自于山采材料。盆栽的主体在大自然的冲蚀下已现雏形，盆景师只需对山采材料进行加工完善。

另一条盆景创作的途径是利用普通的苗圃材料，但是从一粒种子演变为一棵盆景是一个极其漫长的过程。有时你可以在花园中发现原材料，但是这些原材料通常既没有山采材料那般突出的特征，也没有苗圃材料所具有的结构。

然而，马克·诺朗德斯打算向大家证明他能将一株高 5m、粗 50cm 的景观树改作成一盆盆景。面临的第一个挑战便是挖掘出这棵大树并将其装入盆中。让紫杉能够快速地恢复生长，并在 2 年后需要再次移植到盆景盆中。当又老又粗的根死后，许多细根会迅速生长。树干高约 150cm，且不会逐渐变细。活着的分枝能长到 60~100cm 以上。现在面临的问题是如何用这样的材料创作出一盆盆景。

图 5 造型之前，马克正在绘制草图
Before any styling, Marc was drawing a sketch

图 6 用水和塑料刷清理树干
The trunk has been cleaned with water and a plastic brush

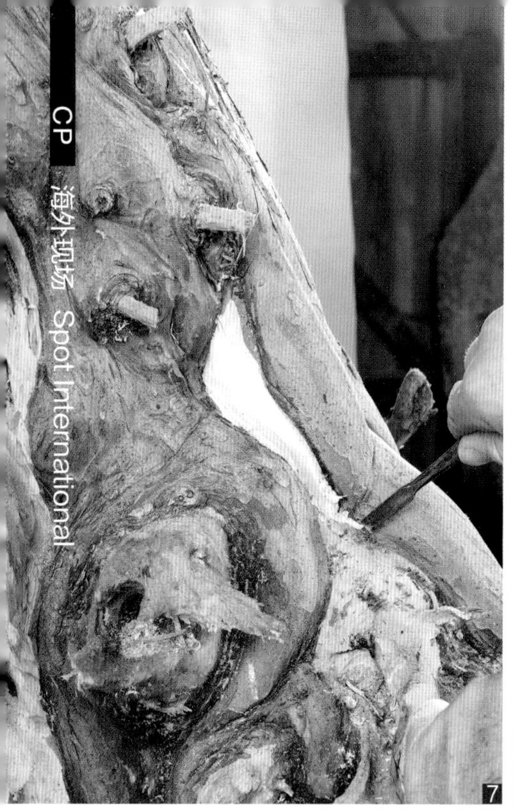

图 7 用刀或刀片移除松散的树皮部分
Loose parts of the bark were removed with a knife or blade
图 8 马克用粉笔标记舍利部分
Marc marked the deadwood parts with a chalk
图 9 用尖凿移除标记的部分
The marked parts were removed with a sharp chisel
图 10 用火烧掉木质纤维。现在你可以决定，舍利的明显程度了
Wood fibers are removed by fire. Now it's your decision, how noticeable you like your deadwood

制作初始

正常情况下，经过造型的盆景才会移植换盆，但在这里有所不同。究其原因是为了找到最好的使根伸展的方向和最佳的观赏面。在不破坏作品的前提下，几乎是不可能将一棵 150cm 以上并做好造型的树移至盆中种植的。根团生长得很好，因此没必要再对其进行雕刻。

树干逐渐变细，四周的根生长情况并不乐观。为了解决这个问题，马克决定将种植在盆中的树拔高，从树桩基部开始雕刻，使得一些细根的表面呈现出已经枯死的形态。

为了解决坏死树干逐渐变细这个问题，他决定用电动工具挖空树干，以便得到更好的锥度。因此，他挖空了树干的一侧，而另一侧则未受影响。

马克很快地选择了正面。主根、右侧用来制作舍利的坏死枝及后方左侧对其成活有利的叶子。恰到好处的分枝对于移植和种植深度至关重要。结合观众的喜好，这将成为唯一的正面。

The beginning

Normally you plant a tree into a pot after the styling work, but in this case there are different reasons to do it the other way round. The main reason was, that he had to find the best nebari and so the best front of the tree. It is almost impossible to repot a just styled tree of more than 150cm without destroying and damaging the work. The root-ball developed well, so there was no need to begin with the carving works.

The trunk tapering and the nebari were very poor. To solve this problems Marc decided to plant the tree higher in the pot, to carve some stumps at the base, so that some of the surface roots appeared as if they were dead.

To solve the problem with the bad trunk tapering, he decided to hollow the trunk with power tools, to get a much better taper. For this reason he hollowed one side of the trunk and let the other side untouched.

Marc chose the front very quick. The main root and the dead branches are on the right side and the useful foliage at the back left side. For the movement and depth it's very important to get a good branch flow. In combination with the leaning to the viewer this will be the one and only front.

图 11 、图 12 他移除了树干的整个树心
He removed the whole heartwood out of the trunk
图 13 马克用链锯处理粗糙的枯枝，目的在于制作出自然变细的效果
Marc did the rough deadwood styling with a chainsaw. The aim was to create a natural tapering
图 14 链锯刀片可做辊式切料机用
The blade of the chainsaw can be used like a rotary cutter
图 15 雕刻工作完成
The carving works are done

图 16 用辊式切料机做精细加工
The precision work was done with a rotary cutter
图 17 经过简单雕刻之后，初见雏形
After the rough carving works, you can see ugly work traces

舍利的制作

清理过树皮之后，马克用粉笔将舍利部分做上标记。制作简单的舍利干并不是一个好主意，因为它已被人们所厌倦。这棵大树需要一个引人注目的亮点！因此，马克开始了为期一周的舍利制作。首先他移除了树皮；接着开始使用大功率工具，如链锯。他开始用链锯挖深洞。他几次剖开树干将木质部挖掉。然后，他又用辊式切料机使舍利变得更加精美。创作一个和谐并有深度的舍利至关重要，舍利要明暗相间。之后，马克在舍利上涂抹了石硫合剂。现在的舍利看起来白的并不那么自然，但是此时此刻已经可以确定哪些部分需要调整了。为了使舍利光滑，马克用砂纸重新打磨了整个舍利。这样，舍利便完成了。

这棵树仍然太高，于是马克将其缩短了约 50cm。它仍然有足够的材料来制作神枝，以便突出树木顶部。

在舍利制作完成时，可以看到马克设计出了一个匀称和谐的造型。

图 18 他用刷子钻头和砂纸打磨表面，使其光滑
With a brush and sandpaper bit, he smoothed the surface
图 19 使用刷子钻头时，务必戴上防护眼镜
When working with a brush bit, it is very important to wear protect glasses

The deadwood

After cleaning the bark, Marc marked the Shari part with a chalk. Creating a simple saba-miki would be a bad idea, because this would be too boring. This huge tree needs attractive highlights! So Marc started on the deadwood for one week. First he removed the bark. Then he started to work with large power tools, like a chain saw. He began to make deep hollows with the chain saw. Several times he opened the trunk and removed large logs of wood. After that he began to work with a rotary cutter to put more details in the deadwood. It's important to create a harmonic and deep deadwood. You must see the game between light and shadow. After that Marc applied lime sulphur on the deadwood. Now the deadwood looked unnaturally white, but at this moment it was possible to define the parts which have to be adjusted. To smooth the deadwood Marc reworked the whole deadwood with sand paper. Then the Shari is finished for that moment.

The tree was still much too high and Marc shortened the yew at about 50cm. He still had enough material to create a top jin (Ten-jin), that would stick out of the top.

At the end of the deadwood work Marc created a well balanced and harmonized sculpture.

图 20 马克用钻头创造出了更自然的景观
With this bit Marc created a more natural look.
图 21、图 22 用砂纸钻头打磨表面
A bit with sandpaper to smooth the surface
图 23 弯曲粗大分枝前，马克移除了部分树皮
Before bending the thick branches, Marc removed a part of the bark
图 24 马克进行木质的雕刻
Marc was carving a notch into the wood

This huge tree needs attractive highlights!

图 25　将轻型燃料喷射在凹陷处，燃料仅在表面燃烧片刻。但然后你会看到一个深褐色且效果好的舍利
Lighter fuel was sprayed into the hollows. This burns only for a short moment and only at the surface. But now you have a real deep black color and a very good effect into the Shari
图 26　轻型燃料快速燃烧，并不会破坏树木
Lighter fuel burns quickly and will be no problem for the tree
图 27　表面发生变化，生成了预期的颜色
The surface was burned and had the wished color, now

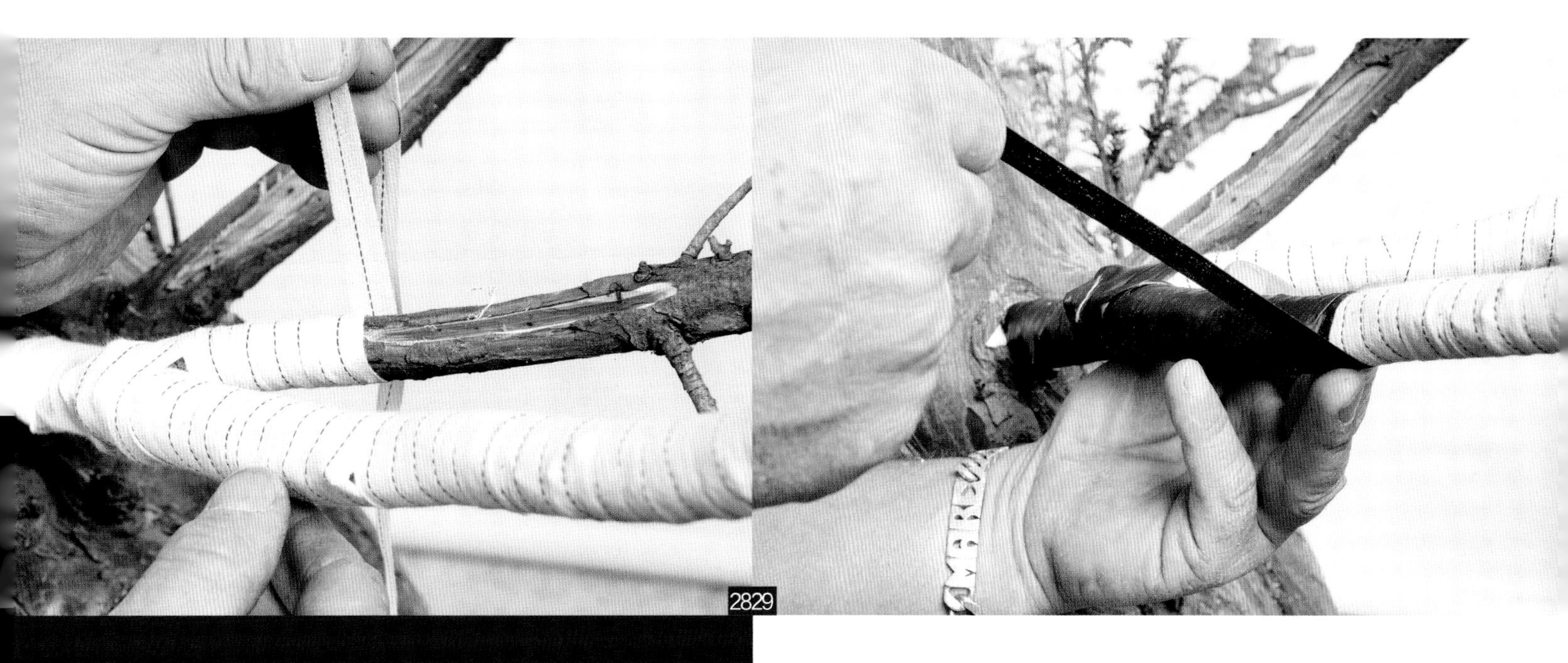

2829

图 28 马克用玻璃纤维胶带(而非棕榈叶纤维)包裹分枝
Instead of raffia, Marc used a fiberglass tape to wrap the branches
图 29 然后在分枝周围包裹上胶带
He also added tape around the branches
图 30 深黑色与右侧舍利形成了鲜明对比
The deep black color forms a very good contrast to the bright deadwood
图 31 重复这些步骤,直至找到想要的结果
Some of the steps can be repeated until you find the wanted result

30

31

树冠造型

现在开始树冠的制作。为了使弯曲时更容易，马克在粗大的树枝上雕刻了一个长20cm、深5mm的刻痕。马克用玻璃纤维布和胶带包裹树枝，以防其在弯曲时受到伤害。接着，他小心地将第一主枝弯曲到相应位置，确定第一主枝的位置后便可以决定树冠其他枝的位置了。

之后，马克开始设计第二个分枝的造型。有些枝交叉在树干中，遮挡了树干的主线，使一棵活着的树的自然感觉呈现出来。一直等到工作结束，马克才确定神枝是要留在树冠正面还是在后方。

图32 舍利制作完成后，开始绕线圈
The deadwood work was done. Now he started to wire the branches
图33 选定分枝后开始缠绕。调整树冠的位置需要许多技能
The chosen branches are wired. Now many skills are needed to position the crown

Styling the crown

Now it was time to create the crown. Therefore Marc made a notch, 20cm long and 5mm deep, at the base of the thick branches. So it was much easier to bend them. To protect the branches, Marc wrapped fiberglass cloth and tape around them. Then he bent the main branch Sashi-eda carefully into position. Now it was possible to decide the position for the rest of the crown.

After this work, Marc began to style the secondary branches. Some of them crossed the trunk and covered parts of this massive trunk line and so supported the natural feeling of a living tree. Marc was waiting till the end of the work to see, if the Jin flows in front or at the rear side of the crown.

图 34 正面。调整树冠
The front side
The corrective work at the crown
图 35 舍利制作完成后，将紫杉移至合适的盆栽盆中种植。根团看起来非常健康
After the deadwood work, the yew was potted in a suitable bonsai pot. The root ball looked very healthy

The future after styling

The main styling will not change very much. The foliage has to become more dense and compact, so the color of the green life veins and the deadwood create a good contrast. The currently used pot suits well, although Marc prefers rectangular pots without edges. The special feature of the styling was the big challenge. Both technical skills as well as creativity were required. The quality of the styling is the result of 30% material and 70% of hard work, in a combination with skill, creativity and perfectionism.

图 36 完成后右侧图
The tree on the right hand side
图 37 完成后左侧图
The tree on the left hand side

造型后期

造型后期，树的主体没有太多变化。叶子变得更加密集紧凑，因此具有生命气息的绿色和白色的舍利形成了鲜明对比。虽然马克更喜欢矩形无边花盆，但现在使用的花盆非常适合这棵树。特别的造型极富挑战性，不仅要求技术技能还要求创造性。造型的好坏取决于30% 原材料所具有的特征和 70% 的辛勤制作，它是技能、创造性和完美主义的集合。

原桩坯 The raw material

The quality of the styling is the result of 30% material and 70% of hard work, in a combination with skill, creativity and perfectionism.

38

图 38 最终结果为高 120cm。马克·诺朗德斯完成了几乎不可能完成的任务。他完成了由一个普通树桩到给人印象深刻的盆景创作
The end result, height 120cm. Marc Noelanders reached the almost impossible. He created an impressive bonsai out of a normal tree stump!

全景报道 2013 比利时第 14 届诺朗德斯杯盆景展

文章来源:《BCI 杂志》Source: *BCI Magazine*
文: 古德·本茨 Author: Gudrun Benz
摄影: 苏放 Photographer: Su Fang

图 1 马克·诺朗德斯宣布颁奖晚宴开始

图 2 诺朗德斯杯展场一角

All Report

2013 Noelanders Trophy XIV

最开始，诺朗德斯杯盆景展的举办地点在赫斯登—佐尔德尔市，位于比利时林堡省，靠近德国西北边境。赫斯登—佐尔德尔市有大约3万人口，大多靠旅游业谋生（主要是徒步旅行和单车骑行）还有做一些看上去并不起眼的小生意。所以，这个地方每年能吸引到这么多的人参观诺朗德斯杯国际盆景展确实很让人惊奇。诺朗德斯杯盆景展这次又换了一个新的地点举办，这次的展览地点放到了已经废弃几十年不用的煤矿大厅。

为了创造出高级别盆景展览的氛围，必须提前对展场进行精心的布置。比利时盆景协会的"工作组"早在展览前一周就开始悬挂非常重的黑色幕布将展场分为中间的展览区以及两边的贸易区。照度很强的聚光灯也已经被固定在天花板上的铁架子上了。除此之外，展台上也已经铺好了蓝色的干净布料。所有这一切精心的准备只是为了在区区两天时间内，让参与者和参观者能得到更好的体验！最后收到的效果是非常不错的。

在2000年，诺朗德斯杯创办之初，在时间上它就成为每年在欧洲最先举行的国际盆景展览。2013年的展览定在1月19日和20日这两天中举行，这又是盆景爱好者们的一次盛世。无论盆景业余爱好者还是专业人士都可以向大家展示他们最好的盆景作品。欧洲的顶级盆景都齐聚在这次非凡的展会上。

这届盆景展有一些小小的变化，因为有众多优秀的盆景作品参加评选，使得诞生了诺朗德斯杯史上最具含金量的两个奖项"最佳盆景（落叶树种）"以及"最佳小品盆景"。每年的一月份，正是落叶树种的整体结构以及分枝结构处于最佳状态的时候。来自西班牙的路易斯·巴列霍先生的山毛榉获得了这届盆景展"最佳盆景（落叶树种）"奖项。

精彩的小品盆景评比展中，来自英国的马克·库珀和瑞塔·库珀脱颖而出，并且连眼光挑剔的参观者也由衷表示赞叹。为了表示对他们的认可，小品盆景一等奖和小品盆景二等奖都颁给了这对夫妇。

现场制作表演方面，来自德国的维尔纳·布希、意大利的恩里科·萨维尼、印度尼西亚的江职宏、美国的苏辛、西班牙的路易斯·巴列霍、荷兰的卡洛斯·范德法特6位盆景大师在1月19日和20日下午进行了现场即兴创作表演。

"诺朗德斯杯"不仅仅是单纯的盆景展览，也是来自世界各地所有爱好盆景的人们的一次大聚会。在这里，人们可以相互交流经验，讨论盆景的未来；在这里，晚宴和颁奖典礼是叙旧和结交新友谊的绝佳时刻。毫不夸张地说，这届展品的平均水平相当高。

展会中一个不可忽视的元素就是贸易区：大片的贸易区有来自英格兰、捷克共和国、德国、法国等全欧洲著名的陶艺家。他们的生意做得相当不错，因为冬天，正是人们打算给他们的树换盆的时候。当然，除了盆景盆之外还有盆景培材等与盆景相关的其他物品也一同在贸易区进行交易。

在1月19日当天，来参观的人群络绎不绝，展区和贸易区被来往人群围的水泄不通。遗憾的是，1月20日天公不作美，一场暴风雪阻止了人们前来参观的热情。

All Report

2013 Noelanders Trophy XIV

Since its beginning, the Noelanders Trophy has taken place in Heusden-Zolder, a municipality in the Belgian province of Limburg near the border of north-west Germany. It has a total population of about 30,000 inhabitants who make their living by tourism (mainly hiking and biking) and some small enterprises—at first glace not very spectacular. Therefore it is amazing how many visitors each year attend the Noelanders Trophy, an international bonsai exhibition, which for the third time was held in a new venue, a large hall of a former coalmine, now closed for several decades.

图 3 印度尼西亚盆景大师江职宏在第 14 届诺朗德斯杯上进行现场表演

图 4 第 14 届诺朗德斯杯盆景展贸易区

In order to create an appropriate atmosphere for a high-level bonsai exhibition, the venue had to be prepared in advance in a carefully considered way. Therefore the "working team" from the Belgian Bonsai Association began one week before the event to hang the partitions of heavy, black-colored fabric that divided the vast hall into the exhibition place in the middle and the vendor area at both sides. Powerful spotlights were fixed to the iron framework of the ceiling. In addition, the rows of display tables had to be set up and covered with clear fabric and blue colored material on the table. And all this for the pleasure of participants and visitors for only two days! But the result was great.

From its beginning in 2000, Noelanders Trophy is always the first international bonsai exhibition of the year in Europe. This year's event took place on January 19 and 20, and was once again a great event for bonsai enthusiasts. Amateurs and professionals alike are allowed to present their best bonsai. Top class bonsai from all over Europe were gathered in this extraordinary show.

There was a small change in the competition policy: For the first time a trophy for the best deciduous tree and a trophy for the best Shohin display were presented as a result of more deciduous bonsai and Shohin compositions than in the past. January is the season when the artistic structure and ramification of a deciduous tree can be seen at its best when no leaves gracefully cover little defaults of the bonsai. The trophy for a deciduous tree went to the outstanding fagus grenata of Luis Vallejo from Spain.

The marvelous Shohin displays of Mark and Ritta Cooper from the UK stood out and were admired by the competent visitors. In recognition, the first and second prize for Shohin went to the couple.

Six bonsai masters gave simultaneous demonstrations on Saturday and Sunday afternoon. Werner Busch, Germany; Enrico Savini, Italy; Robert Steven, Indonesia; Suthin Sukosolvisit, United States of America; Luis Vallejo, Spain; Carlos Van der Vaart, Netherlands.

Noelanders Trophy isn't only a bonsai show but the event is also an international meeting of friends with a common love of bonsai, it is a time for exchanging experiences and discussing the future of the beloved bonsai. Always excellent, the gala dinner and awards ceremony is the right moment to renew friendships or to make new ones. One can say without exaggeration that the general level of the exhibition was also very high this year.

Another element of the event is conspicuous: There is always a large trader area with lots of well-known ceramists from all over Europe. This time, they came for example from England, Czech Republic, Germany, and France. They have done good business – in wintertime, bonsai people are planning the re-potting of their trees. Of course, there were many other stands with bonsai, raw material for bonsai and bonsai related items.

On Saturday, the stream of visitors was endless and the exhibition and trader areas were overcrowded. Unfortunately, the weather changed on Sunday, a snowstorm prevented many people from coming.

图 5 马克·诺朗德斯为路易斯·巴列霍颁发"最佳五叶松提名奖"

图 6 诺朗德斯展场一角

"2013 第14届诺朗德斯杯"盆景展全景记录——颁奖晚会

Panoramic Record of the 2013 Noelanders Trophy XIV —Awards Ceremony

盆景·热情·交流·畅想

文: CP 摄影: 苏放
Author: CP Photographer: Su Fang

"诺朗德斯杯"的举办时间正是最寒冷的冬天，但是颁奖晚会上的热烈气氛把严寒驱散，人们对盆景的热爱热情期盼彻底释放宣泄出来了。交流技艺，畅想未来，使人们从未对盆景的未来有如此多的想象。

图1 比利时盆景协会会长马克·诺朗德斯先生为盆景制作表演者路易斯·巴列霍先生（西班牙）颁发纪念品
Mister Luis Vallejo, Spain is receiving a souvenir from Mister Marc Noelanders

图2 比利时盆景协会会长马克·诺朗德斯先生为盆景制作表演者苏辛先生（美国）颁发纪念品
Mister Suthin Sukosolvisit, U.S.A is receiving a souvenir from Mister Marc Noelanders

图3 盆景制作表演者恩里科·萨维尼先生（意大利）领取纪念品
Mister Enrico Savini , Italy is receiving a souvenir

图4 盆景制作表演者维尔纳布希先生(德国)领取纪念品
Mister Werner M. Busch, Germany is receiving a souvenir

图 6 欧洲盆景协会会长格雷格·博尔顿
Mister Greg Bolton, chairman of European Bonsai Association

图 7 加西亚·费尔南德斯·艾拉斯莫先生代马里亚诺·爱莎先生领取欧洲盆景协会优秀奖
Mister Garcia Fernandez Erasmo is receiving Award of Merit EBA for Mister Mariano Aisa

图 5 盆景制作表演者卡洛斯·范德法特先生(荷兰)领取纪念品
Mister Carlos Van Der Vaart, Netherlands is receiving a souvenir

图 8 来自国际盆景协会的凯斯·休斯女士
Miss Kath Hughes from International Bonsai Club

图 9 凯斯·休斯女士为马克·库珀和瑞塔·库珀颁发国际盆景俱乐部优秀奖
Mister and Mrs. Mark and Ritta Cooper is receiving Excellence Award of Bonsai Club International given by Miss Kath Hughes

图 10 凯斯·休斯女士为莫罗·斯丹姆伯格先生颁发国际盆景俱乐部优秀奖
Mister Mauro Stemberger is receiving Excellence Award of Bonsai Club International given by Miss Kath Hughes from BCI

图11托尼缇科先生代马里亚诺爱莎先生从维尔纳布希先生手中领取杜塞尔多夫盆栽美术馆特别奖
Mister Werner M. Busch (left) is handing over the Special price from Bonsai Museum Düsseldorf. Mister Tony Tickle is receiving it for his friend Mister Mariano Aisa

图 12 克里斯蒂安·沃斯先生领取比利时盆栽协会最佳参与奖
Mister Christian Vos is receiving "Best participating BAB"

图 13 伍德·费舍尔先生的鸡爪枫盆景获得提名奖
Mister Udo Fisher is receiving "Nomination for *Acer palmatum* Bonsai"

图 14 路易斯·巴列霍先生领取"最佳盆景(落叶树种)"奖项
Mister Luis Vallejo is receiving"Best deciduous Bonsai"

图 15 南希·威丽女士领取“最佳小品盆景提名奖”
Miss Nancy Vaele is receiving “Nomination for Shohin”

图 16 马克·库珀和瑞塔·库珀领取“最佳小品盆景奖”
Mister and Mrs. Mark and Ritta Cooper is receiving “First price Shohin”

图 17 弗雷德里克·夏奈尔先生的石榴树盆景获得提名奖
Mister Frédéric Chenal is receiving“Nomination with a *Punica granatum* Bonsai”

图 18 加西亚·费尔南德斯·艾拉斯莫先生的油橄榄盆景获得提名奖
Mister Garcia Fernandez Erasmo is receiving “Nomination for *Olea europea sylvestri* Bonsai”

图 19 路易斯·巴列霍先生的五叶松获得提名奖
Mister Luis Vallejo is receiving“Nomination for *Pinus pentaphylla* Bonsai”

图 20 路易斯·巴利尼诺先生的鱼鳞云杉获得一等奖
Mister Luis Balino won the first price for *Picea jezoensis* Bonsai

路易斯·巴列霍在第14届诺朗德斯杯上的现场表演

Luis Vallejo's Demonstration on Noelanders Trophy XIV

摄影: 苏放 Photographer: Su Fang

1

原桩坯

Luis Vallejo's

2

3

4

5

原桩坯

6

完成后

1 原桩坯

2

卡洛斯·范德法特 在第 14 届诺朗德斯杯上的现场表演

Carlos Van der vaart's

Demonstration on Noelanders Trophy XIV

摄影: 苏放 Photographer: Su Fang

NOELANDERS TROFEE XIV
styled by :
Carlos
Van der vaart

完成后

原桩坯

苏辛 在第 14 届诺朗德斯杯上的现场表演

Suthin Sukosolvisit's Demonstration on Noelanders Trophy XIV

摄影: 苏放 Photographer: Su Fang

1 原桩柸正面

2 原桩坯背面

3

原桩坯

完成后

原桩坯

恩里科·萨维尼
在第 14 届诺朗德斯杯上的现场表演（一）

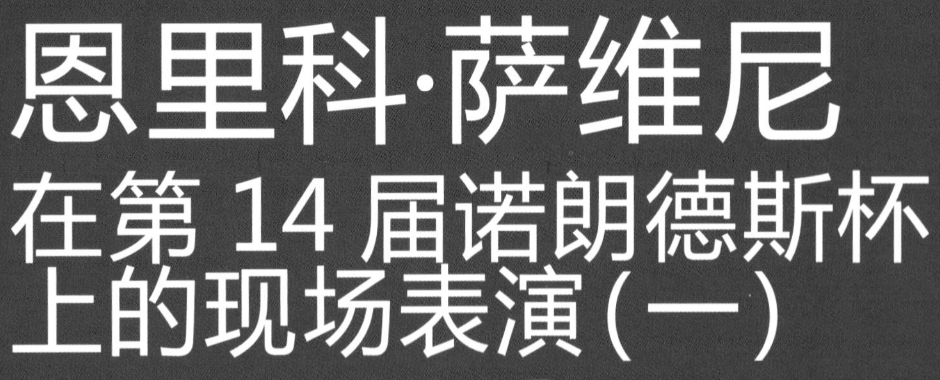

Enrico Savini's
Demonstration on Noelanders Trophy XIV (1)

摄影：苏放 Photographer: Su Fang

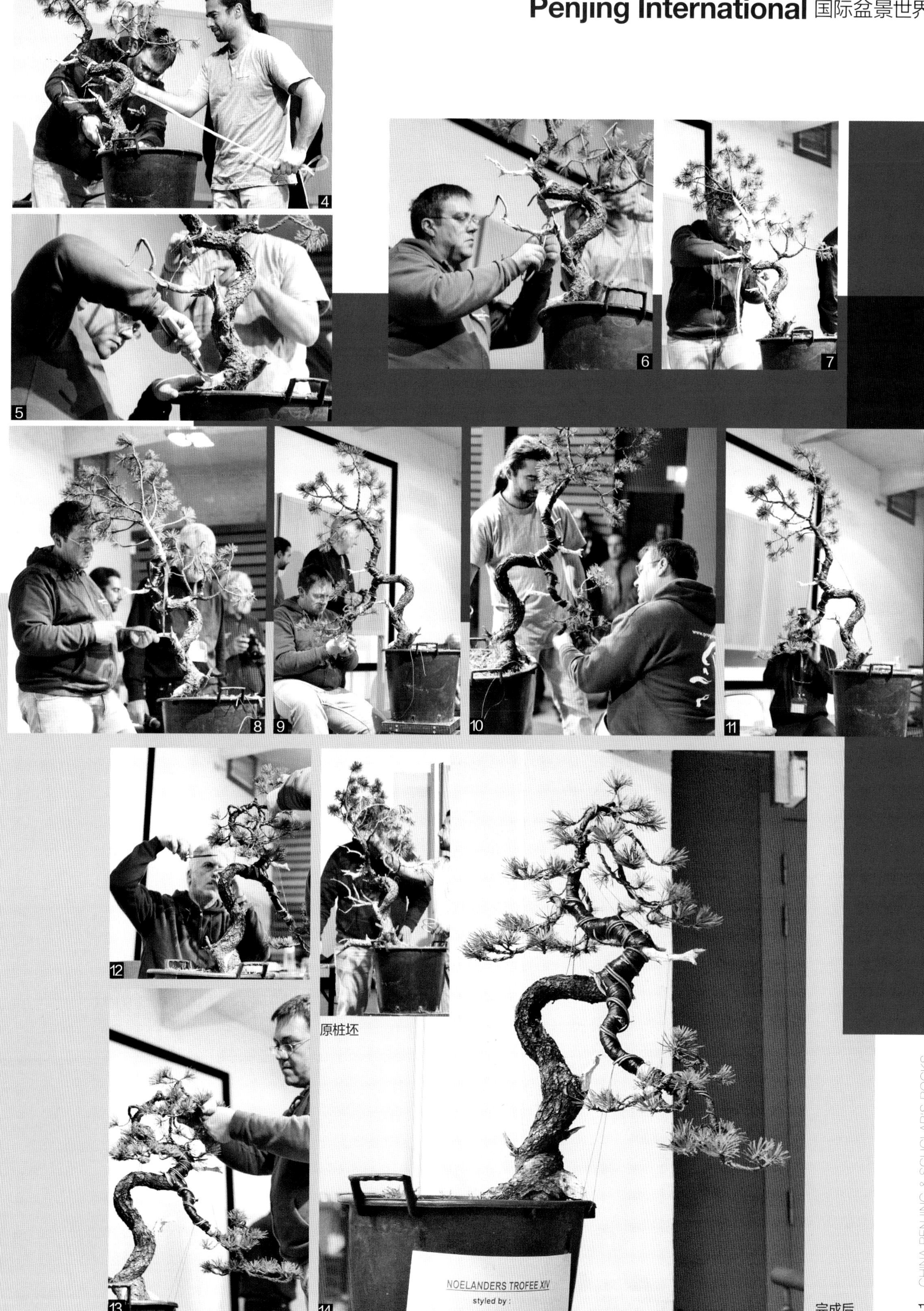
4
5
6
7
8
9
10
11
12
13
14
NOELANDERS TROFEE XIV
styled by :

原桩坯

完成后

恩里科·萨维尼
在第 14 届诺朗德斯杯上的现场表演(二)

Enrico Savini's Demonstration on Noelanders Trophy XIV (2)

摄影: 苏放 Photographer: Su Fang

1 原桩坯

6

7

8

9

10

完成后

原桩坯

11

维尔纳·布希 在第 14 届诺朗德斯杯上的现场表演（一）

Werner Busch's Demonstration on Noelanders Trophy XIV (1)

摄影: 苏放 Photographer: Su Fang

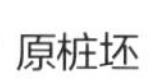
原桩坯

原桩坯

维尔纳·布希在第14届诺朗德斯杯上的现场表演（二）

Werner Busch's Demonstration on Noelanders Trophy XIV (2)

摄影: 苏放 Photographer: Su Fang

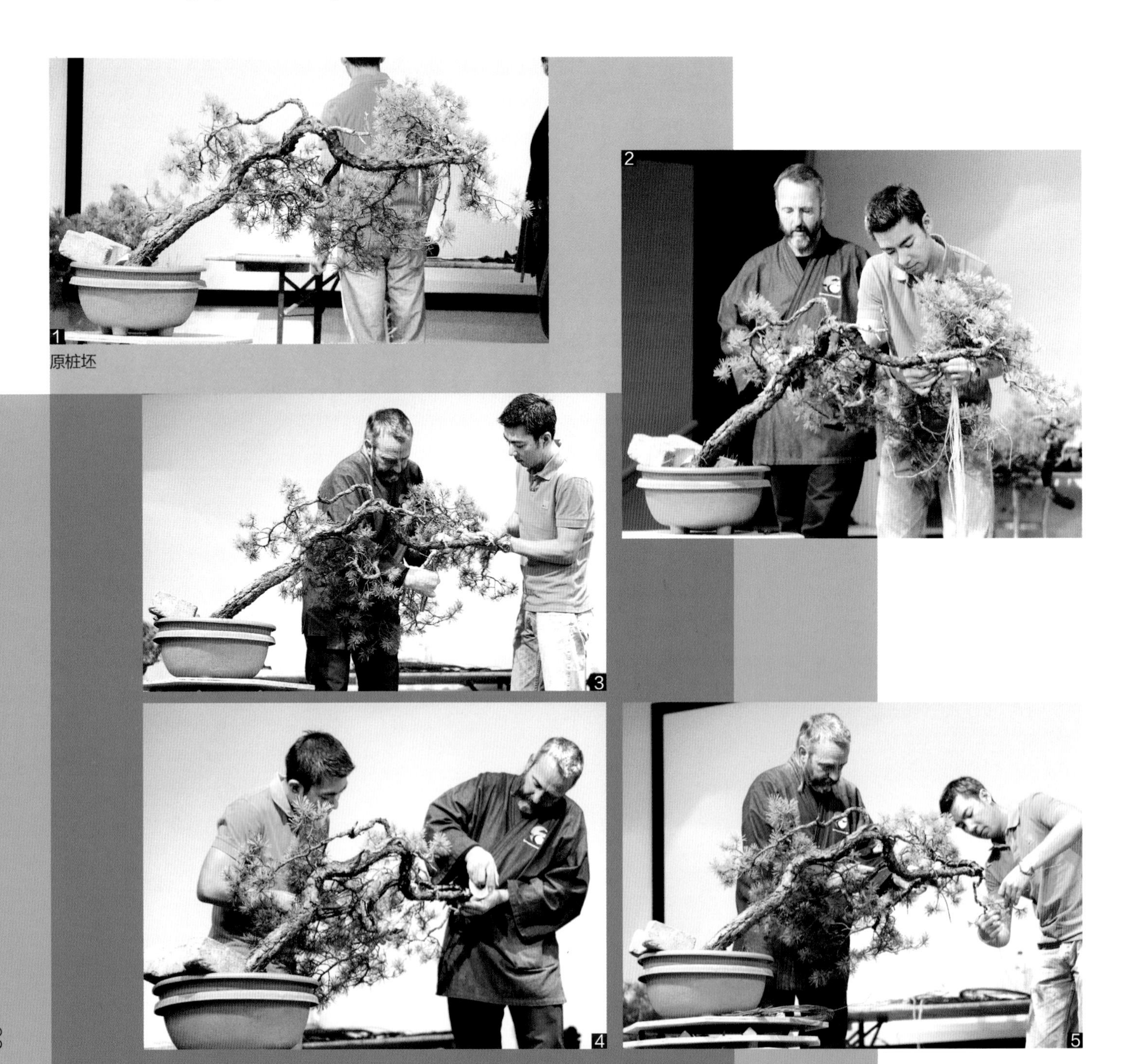

原桩坯

原桩坯

完成后

江职宏
在第 14 届诺朗德斯杯上的现场表演（一）

Robert Steven's
Demonstration on Noelanders Trophy XIV (1)

摄影: 苏放 Photographer: Su Fang

1 原桩坯正面
2 原桩坯俯视图
3
4

5
6
7

8 9 10 11

12

13

14
15

NOELANDERS TROFEE XIV
styled by :
Robert
Steven
16

原桩坯

完成后

江职宏 在第 14 届诺朗德斯杯上的现场表演（二）

Robert Steven's Demonstration on Noelanders Trophy XIV (2)

摄影: 苏放 Photographer: Su Fang

1

2

原桩坯

3

4

5

6

7

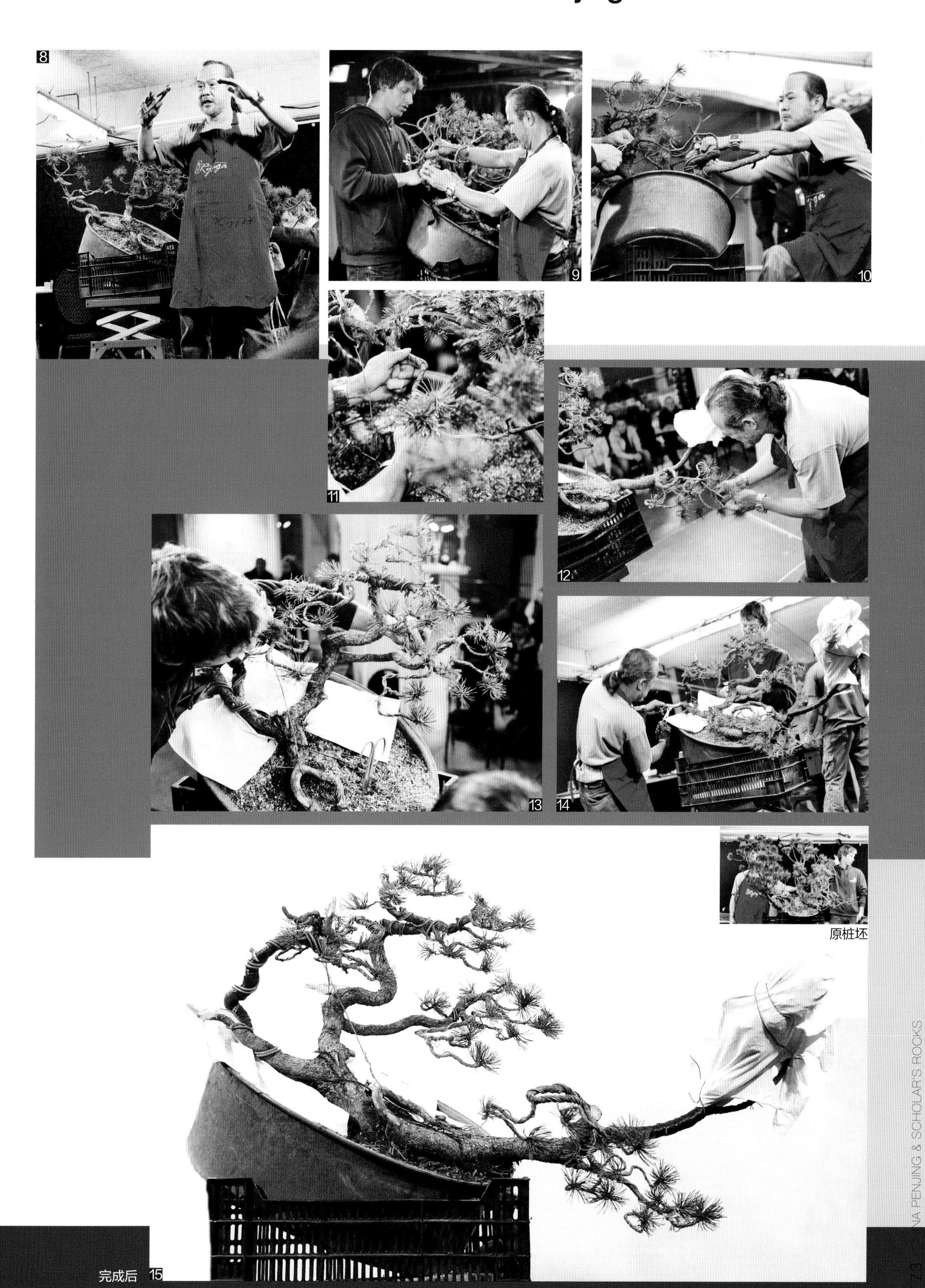

原桩坯

完成后 15

第14届诺朗德斯杯

作品全景

摄影: 苏放 Photographer: Su Fang

Works Appreciation of Noelanders Trophy XIV

第14届诺朗德斯杯

作品全景

摄影: 苏放 Photographer: Su Fang

Works Appreciation of Noelanders Trophy XIV

第14届诺朗德斯杯

作品全景

摄影: 苏放 Photographer: Su Fang

Works Appreciation of Noelanders Trophy XIV

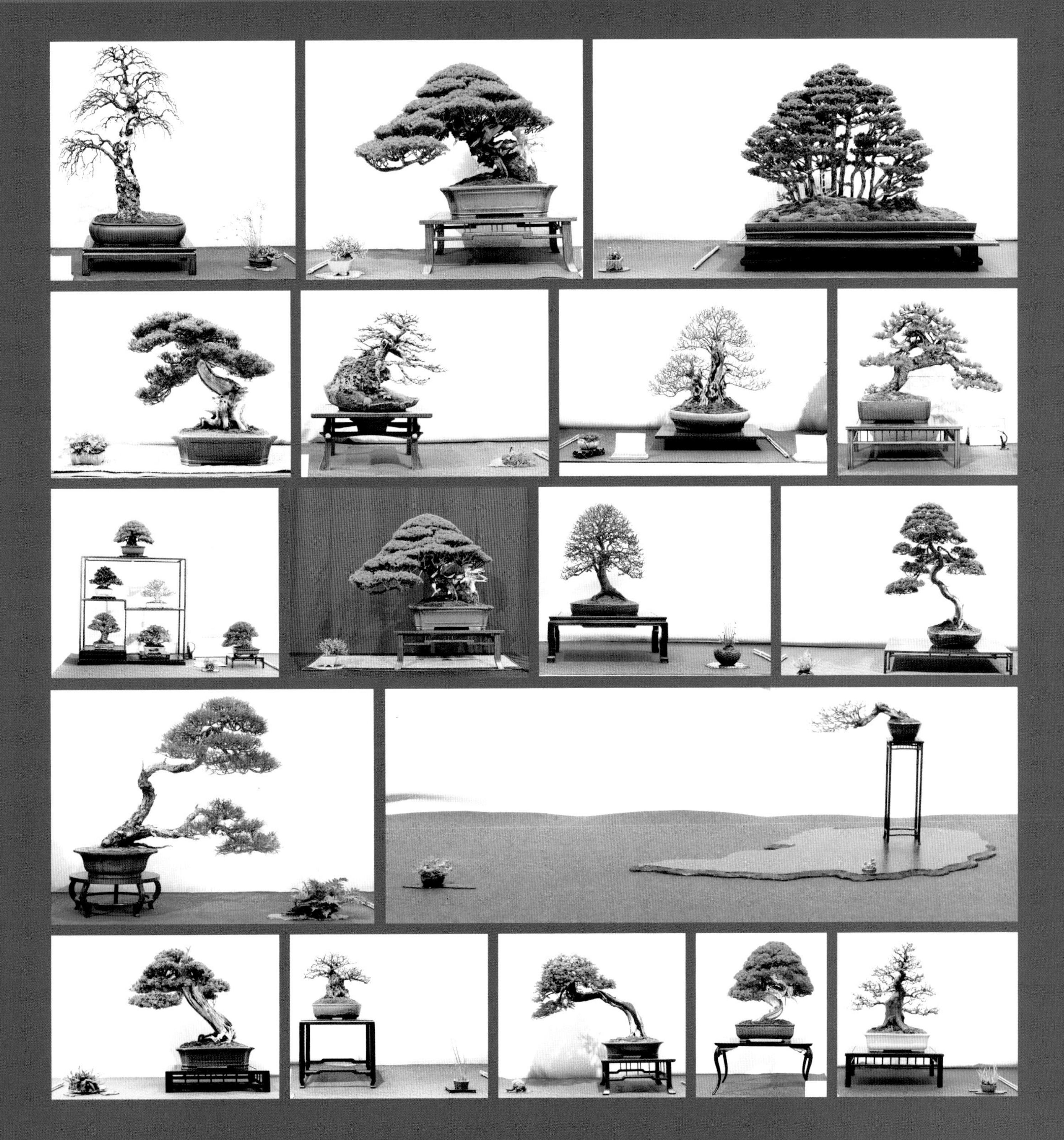

巴登符腾堡
Castle of Bad Rappenau

第四届杜鹃花节，“日本花之梦”

2013 年 6 月 1 日 ~2 日举办于德国巴登符腾堡水上古堡

4th Azalea Festival

“Japanese Blossom Dreams”

at the Moated Castle of Bad Rappenau, Germany from June 1 ~ 2, 2013

文、供图: 古德·本茨　Text/Photos: Gudrun Benz

巴登符腾堡
Castle of Bad Rappenau

为期两天的展览举办于巴登符腾堡州，巴登符腾堡州是一个面积不大的州，位于德国南部。巴登符腾堡州居住有20000居民。它的历史可以追溯到14世纪。19世纪盐水的发现使这个州逐渐发展成为一个温泉小镇，20世纪末慢慢又出现了许多供人疗养的地方。

巴登符腾堡（见图），是这个小镇的地标性建筑，在这浪漫的地方任何梦想都可以成真。所以，这是举办“日本花之梦”杜鹃花节的绝佳之处。这次活动的组织者伍德·费舍尔先生是德国最著名的盆景制作表演者之一。

展览的内部视图
View into the exhibition

展览的内部视图
View into the exhibition

壁龛内陈列着李斯特·韦勒夫人的杜鹃花品种的盆景藏品
日本盆
Tokonoma display with a bonsai tree of Mrs. Liselotte Weller, Pot: Japan

壁龛内陈列着伍德·费舍尔先生的杜鹃花盆景藏品 日本常滑烧盆
Tokonoma display of Mr. Udo Fischer Satsuki azalea variety Kakuo, pot Tokoname

近距离观赏塚田博巳先生的盆景藏品
Close view at the azalea (rhododendron indicum 'Kinsai') of Mr. Hiromi Tsukada

塚田先生盆景藏品的细节展示图
Detail of the azalea bonsai of Mr. Tsukada

壁龛内陈列着塚田博巳先生的盆栽藏品
Tokonoma display with a tree of Mr. Hiromi Tsukada

杜鹃花 日本盆 收藏者:齐格弗里德范德先生藏品
Satsuki Azalea, variety: Komane, pot: Japan, collection Mr. Siegfried Vendt

来自日本东京的塚田博巳是一位盆景制作表演者，同时也是此次活动的贵宾。他是一位有名的盆栽大师，在日本举办的展览中得到了许多奖项。他骄傲地指出他的所有树都是从种子或是修剪一点一点亲自完成的。在展览之前，塚田博巳已经举办了一整天关于制作养护杜鹃花的培训课程。

展览安排在建筑的第一层举行。虽然是一个规模不大的展览，但是盛开杜鹃花的美妙景象会久久地停留在参观者的回忆中。

The two day exhibition took place at Bad Rappenau, a small town situated in southern Germany. Bad Rappenau is a community with a population of about 20000 inhabitants. Its history goes back to the 14th century. When a brine was discovered in the 19th century it slowly developed in a spa town where in the late 20th century several rehabilitation clinics were added.

The moared castle (Figure) is the landmark of the small town and a romantic place where dreams can become true. So it is an ideal location for an azalea festival named "Japanese Blossom Dreams". The event was organized by Mr. Udo Fischer, one of the best bonsai demonstrators in Germany.

Mr. Hiromi Tsukada from Tokyo, Japan was a demonstrator and a guest of honour of the event. He is a well known bonsai master who won several prizes at exhibitions in Japan. He proudly points out that all his trees are elevated from the seed or cuttings. Already before the exhibition Mr. Tsukada gave a whole day workshop about creating and training azalea bonsai.

The exhibition was established on the first floor of the building. Even if it was only a small exhibition it was very impressive by the splendour of the azaleas in full blossom which beauty will last in the memory of visitors for a long time.

杜鹃花 日本常滑烧盆 收藏者:伍德·费舍尔藏品
Satsuki azalea variety Seidai, pot: Tokoname, collector: Mr. Udo Fischer

杜鹃花 日本盆 约翰·卡斯特纳藏品
Satsuki azalea variety Matsuno Homare, pot: Japan, collector: Mr. Johann Kastner

杜鹃花 日本常滑烧盆 沃尔夫冈·加特纳藏品
Satsuki azalea varieta Gyoten, pot: Tokoname, collector: Mr. Wolfgang Gärtner

杜鹃花 日本常滑烧盆卡尔·厄尔本藏品
Satsuki azalea variety Katsusamurasaki, pot: Tokoname, collector: Mr. Karl Erben

杜鹃花 彼得·克雷布斯盆 厄休拉凡克藏品
Satsuki azalea variety Gyoten, pot: Peter Krebs, collector: Ursula Funke

杜鹃花 日本常滑烧盆 迈克尔·赫林格尔藏品
Satsuki azaleaVariety Kazan, pot: Tokoname, collector: Michael Herrlinger

杜鹃花 日本盆 珍妮·德罗斯特藏品
Satsuki azalea variety Nyohasan, pot: Japan, collector: Mrs. Janine Droste

杜鹃 收藏者：李斯特·韦勒藏品
Satsuki azalea, collector: Mrs. Liselotte Weller

杜鹃 收藏者：李斯特·韦勒藏品
Satsuki azalea, collector: Mrs. Liselotte Weller

杜鹃 收藏者：李斯特·韦勒藏品
Satsuki azalea, collector: Mrs. Liselotte Weller

杜鹃花 日本盆 收藏者：约翰·卡斯特纳藏品
Satsuki azalea variety Matsu no Homare, pot: Japan, collector: Mr. Johann Kastner

塚田博巳先生
Mr. Hiromi Tsukada

海峡两岸名家书画暨耐阴茶壶盆景展于 2013 年 7 月 26 日至 8 月 26 日在深圳东湖美术馆举行

开幕式现场

出席开幕式的嘉宾 右二中国国际书画家收藏家协会会长黄杰力、左二林鸿鑫、左三陈习之

Cross-straits' Painting and Calligraphy & Teapot Penjing Exhibition Will Be Held in Shenzhen Donghu Art Museum during July 26th to August 26th,2013

2013年7月26日在深圳东湖美术馆，由深圳市风景园林协会盆景赏石文化分会及中国国际书画家收藏家协会联合主办，深圳市宏星园林景观有限公司承办的“海峡两岸名家书画暨耐阴茶壶盆景展”隆重开幕。

本次展览作品由刘大为、韩美林、刘小刚、程熙、藏敬儒等著名书画家及中国盆景艺术大师林鸿鑫夫妇提供。

盆景被视为无声的诗、立体的画，而在人们传承千年的茶壶里配上精美的盆景更属珍品，它与名人字画碰撞出的艺术火花更是灿烂无比。本次展览是一次书画艺术与盆景艺术的碰撞与结合，开启了书画与盆景合璧联展的先河，是中国文化在书法、绘画、盆景中互相交融与发展的诠释，是中国盆景逐步复兴并和其他艺术一同展现中国风格魅力的起步，是创新理念下一种具有文人特色的盆景形式登上舞台，也使得盆景可以作为高雅艺术在美术馆中展览这一中国盆景人的愿望实现。

展期，栩栩如生的绘画、苍劲有力的书法、生机勃勃的盆景共放光彩，吸引了广大市民，并对此赞不绝口，给盛夏的鹏城带来了一丝凉意，也让今年的盆景盛会更多彩！

山东省盆景艺术家协会第四届会员代表大会于2013年7月23日召开

The 4th Members' Congress of Shandong Province Penjing Artists Association on July 23th, 2013

供稿：武广升、卢宪辉 Source: Wu Guangsheng, Lu Xianhui

2013年7月23日上午，山东省盆景艺术家协会第四届会员代表大会在济南桃花岛人和书院召开。经山东省林业厅同意，山东省民政厅批复，山东省盆景艺术家协会进行换届选举，来自各市、县、区的百余名会员表决通过，选举新一届领导班子。

山东省政协原副主席王宗廉及省民政厅领导到会祝贺，并在讲话中说到："展望新一届工作，山东省盆景艺术家协会定会在党的十八大精神指引下，继续遵循'保护山林，栽培小苗'的目标，以'讲团结、讲艺德、讲奉献、讲创新'为宗旨，使山东盆景发展再上一个新台阶！"会议期间，新任会长田庆彬也做了相应发言。

本次会议通过了山东省盆景艺术家协会第四届领导机构，具体名单如下：

名誉主席：山东省原副省长王裕晏，山东省政协原副主席王宗廉；

名誉会长：中国盆景艺术家协会名誉副会长魏绪珊，中国盆景艺术家协会副秘书长吉敦训；

会长：田庆彬；

副会长：孙景海、马建奎、付聿胜、崔周村、黄士东、孙伯文、齐德武、武广升；

秘书长（兼）：付聿胜。

会期，山东省盆景艺术家协会公布成立了3个专业委员会和1个书画院，分别为：盆景委员会、奇石委员会及根雕委员会和盆景书画院。同时，大会还公布了如临沂奇园、青州知松园、日照魏园等20个盆景名园及其他事项。

本次山东省盆景艺术家协会会员代表大会在全体领导和与会会员的共同努力下，圆满结束。

华南地区最大的绿化苗木交易中心 —— 南方绿博园

The Largest Green Seedings Trading Center in South China Area –South Green Garden

盆景园指示牌

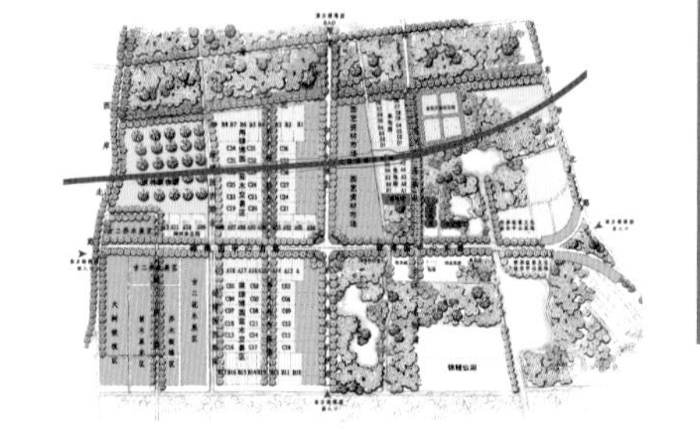
总平面图

盆景园效果图

南方绿博园 —— 华南地区最大的绿化苗木交易市场之一，所打造的平台不仅是一条以销促产、产销结合、工程服务为一体的产业链，同时也是一个集展示交易、科普教育、休闲旅游的生态式综合园区；不仅是苗木产业的经济助推器，更是培养绿色文化、提升绿色品位的生态孵化器。

南方绿博园概况

中山市南方绿博园（简称"绿博园"）位于中山、江门、佛山三市交汇处的古镇镇，毗邻港澳，交通便利，占地 10050 亩，自 2007 年至今举办了 4 届中国（中山）南方绿化苗木博览会，经过 5 年的规划建设，各个展览区和各项配套设施日趋完善，逐步发展成以绿化苗木产业为主导，配套盆景奇石、高端锦鲤、珍稀龟鳖养殖等产业并驾齐驱的综合性产业园区，以其科学的规划和合理的布局得到了社会各界的一致好评和高度认同。

万亩南方绿博园 打造生态绿色经济活力舞台

园区设有多个展区：现有 8000 亩苗木交易区、7800m^2 的苗木主展馆、苗木新品种展示区、100 亩奇石盆景园、两期 200 亩盆景精品交易园区、200 亩名龟园交易市场、140 亩龟鳖养殖示范区、350 亩观赏鱼养殖区、观赏鱼检测检疫中心、锦龙皇府（锦鲤展区）等和正规划建设 200 亩园艺资材及园林用品专业市场。

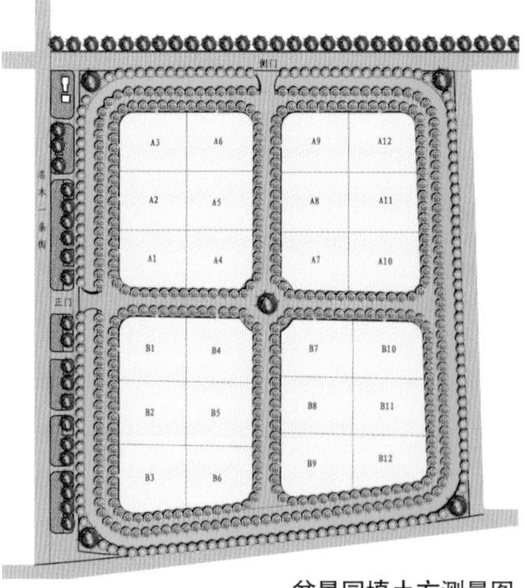
盆景园填土方测量图

工农并举，工业反哺农业 中山苗木向周边地区延伸

不少人知道古镇以灯饰闻名，古镇镇乃实至名归的"中国灯饰之都"，却不知古镇的苗木也有着 40 多年的历史。如今，古镇的灯饰产业已经在周边镇区形成了完善的产业链。然而，绿化苗木也是当地的另一标志产业，在 2011 年，古镇镇被评为"中国花木之乡"。以古镇镇为起点，向南延伸的中山花木交易市场早已成为华南地区最大的绿化苗木交易市场。也引领着国内的苗木绿色行业的发展。

南方绿博园盆景精品交易园区（简称：盆景精品交易园），位于南方绿博园内，与新规划的江珠高速北沿线古镇出入口交接，交通便利。整个项目规划面积约 200 亩，第一期规划面积约 100 亩，园区外围以河流环绕封闭，周边种植木棉和黄花风铃木等开花树种，环境优美。盆景交易园设两个主要出入口，以半封闭式物业管理，安全可靠，先推出一期共划分 24 块作为盆景艺术交易区。二期将于明年推出。

南方绿博园盆景交易园区建成后将是一个高规格的专业盆景交易园，引进既有较高盆景创作艺术水平，又活跃于市场交易的创作者和企业进驻交易园区，展销较高水平的盆景精品，汇聚广大热爱盆景的爱好者在此进行交流和交易，逐步形成一个面向华南地区的盆景交易专业市场和盆景艺术文化创作推广中心。今后还将继续与中国盆景艺术家协会、广东省盆景协会不断推动盆景精品交易平台建设，加快盆景产业文化发展，定期开展周期性盆景拍卖活动，逐步形成一个面向华南地区的盆景专业交易市场，为推动中国盆景文化走向世界作出一份使命性的奉献。

古镇人挑灯走天涯，以一盏灯在工业领域书写了古镇传奇。同样，古镇人又围树而坐，笑迎客来。可以预见的是，将来的古镇人，将以绿树锦鲤名龟，在广泛农业领域再次创造奇迹。

注：一亩 =1/15hm^2=666.67m^2

一生的选择

The Choice of a Lifetime

文：张桂庆 Author: Zhang Guiqing

“崖上古柏”

我是一个47岁的农民，从事盆景种植创作近30个年头，有一个近万平方米的专业盆景园。2002年8月由张世藩老师和一个朋友推荐加入中国盆景艺术家协会，10年来随着我国经济的迅猛发展和协会的栽培，我已成为“中国盆景艺术家协会”理事、河南省中州盆景艺术大师、河南省项城市盆景协会会长……从一个农民到大师，从一个会员到领导上百个会员的会长，这标志着我国盆景事业在组织的领导下所看到的进步和辉煌成绩。当然这里也有朋友的支持和我的努力，最应该感谢的还是“中国盆景艺术家协会”。

我们的成功做法是：因地制宜，培育桩材

因为我们地处中原，树桩资源贫乏，大部分树桩都是自己培育的，用我的经历让爱好者看到从一颗小苗能培养成价值上千元的优秀作品，使每个爱好者都能接受盆景。

共同学习，共同提高

我们一些有成熟经验的盆景人，不要认为自己是大师，是会长……我们也是学员，我们要把爱好者当成朋友，互相交流，共同创作，互帮互学，共同提高。

吸收会员，建立组织

我们通过创作表演，盆景展示，提高群众热爱盆景的兴趣，把自己所学教给他们，然后建立组织，吸收他们为会员。在组织的引导下定期交流、学习、参观、参展，使更多爱好者向组织者靠拢。

发挥组织优势，做到推进发展

通过组织共同学习、提高、交流、参观，展示会员作品，推销会员作品。然后引进兄弟单位新枝法、新品种，搞好地方盆景展示和企业单位赞助展览，使会员们不花钱就能玩盆景。

近些年由于我们做到了以上几点，我们的会员来自各个方面，同时也受到了政府、机关、学校、企业的关注，我想我们中国盆景人能做到无私，不高调地去交朋友，去教学员，我们的人生更精彩，能把中国盆景发展成中国文化，我们的人生将更有价值!

“翠”

“冲上云霄”

1 2

论盆景360°全景展示

Talk About 360 Panor

—The Creation Technology of Yaccatree Penjing "Green"

文：关山 Author: Guan Shan

本文通过介绍了360°全景展示技术在盆景中的有效实现，探讨360°盆景展示技术概念、特点、关键技术等，从而推进盆景技术的可持续性创作及业态发展。

盆景是我国传统文化艺术，历史悠久、源远流长，它以饱含诗情画意、讲求神韵和意境的艺术特色，闻名于世。从出现盆景雏形时期发展至今，一直分有流派和形式，就这两者一直探讨至今，但鲜见跳出两者外以第三者概念论述盆景。经常有人欣赏盆景是先看"面"，这不是眼界限制，而是欣赏习惯，让我们释怀"心界"改变这种长久习惯，用360°全景概念来品析罗汉松盆景新作——"苍翠"。

为什么盆景要有它固有的某种形态,非得给它设定一个欣赏角度吗?未必,因为盆景产业的特殊性在于它不是一个依靠生产线与习惯来控制而存在的行业,同时专业领域及专业人才的"细分"催生出盆景新的存在形式。局限角度只是个别人的欣赏界限问题,好的盆景应该从多角度去欣赏。就这一概念,我暂且将之定为"盆景360°全景展示技术"。

盆景是一种集合创作思维、协同各方面技术、整合现有素材而存在的一种美学行为。其创作思维的先进性可大大提高盆景创作的多样性,同时改变盆景素材的欣赏价值。

环顾罗汉松盆景"苍翠",其气魄宏大、富有强烈的艺术感染力、作品结构严谨、古朴自然、各单元之间具有十分清晰的内在联系,使各角度具备独特看点,同时兼顾了盆景整体的融合性,以达到360°全景展示的效果(见图1~4)。要具备360°全景展示效果,对素材的培养尤为重要,素材最好八方出根,树身挺拔,出枝必须四面环顾,制作上要强调各大小单元之间强烈的逻辑关系,在空间上刻画出360°的立体画面。一张照片并不能够充分体现出盆景的韵味,照片只是对作品的一个平面表述,而盆景是放置于庭院的立体作品,应具备360°观赏效果。

随着中国盆景业的飞速发展,素材所包含的种类也越来越多,所能表现出的效果也越来越好,而一些比较传统的表现方式也越来越无法满足盆友们的欣赏要求。在传统的表现方式中,展示的手段无非是定面选角,这无疑不能有效全面地展现盆景;而360°全景展示技术虽然不能用于所有素材,但这一概念可以更好地提升我们对盆景的多方位创作。所以,当我们需要全面表现盆景时,360°全景展示技术无疑是最好的制作选择。

生活才是重点,工作仅是附属品,一个盆景从业者,若没有生活的阅历,不能洞察生活细节,怎么可能做出符合当下的盆景艺术呢?好盆景,靠的是扎扎实实的专业技术和对生活的领悟。

经过这些年来的生活体验,我逐步认识到人生确实是一个过程,无论你在这个过程中扮演什么样的角色,做自己、做自己觉得有价值的事情并将之变成自己的事业,这个很重要。于是,我决定一直朝着这个方向,努力去体验我想要的生活状态。

ic Display of Penjing

3

4

“向上”对节白蜡 *Fraxinus hupehensis* 高 90cm 宽 70cm 刘永辉藏品 苏放摄影
“Up-right”. Height: 90cm, Width: 70cm. Collector: Liu Yonghui, Photographer: Su Fang

点评 Comments

“向上” 对节白蜡

Fraxinus hupehensis “Up-right”

文：李奕祺 Author: Li Yiqi

盆中十来棵树，主干基部环抱一块巨石向上生长。中央一棵最高，成为主心骨，其他分布两旁，依高矮顺序排列，呈正三角形顶。

各主干并非如竹筒子笔直向上，而是各自在小空域中弯角逆回转折，浑力雄劲，犹有一股无形的热气流蒸腾驱动，枝干以飘动的形态放纵向上，争高斗昂，升势劲起，气脉喷张。树干嶙峋洒脱，伸中有缩，缩中萌伸，拐弯处肿胀及疤痕，突显年功老到，蕴含有骨、肉、筋、脉的力量，树枝丰满柔韧，刚劲与委婉交织，表现出轻、重、缓、急的线条变化，彰显出欣欣向荣，蒸蒸日上，节奏明快，韵律流畅，线条动感上飘的画面。

线条美是中华民族特殊的审美观：认为线条无特定内涵，而线条又是内涵极丰富的艺术表现形式，有无限的包容性，可寄情达意，表现丰富的内心世界。线条美被广泛用于中国汉字书写、水墨画及盆景的创作等领域。感悟其中蕴含的轻重力量变化的美妙，从快慢的韵律中产生美感的艺术升华。线条美只有通过感悟才能达到较高的虚幻玄妙境界，而较难用平实的言语去圆满表达。这就是欣赏线条美的妙不可言的艺术境界。

此盆“对节白蜡”似火盆中燃烧着永不熄灭的熊熊烈焰，整体枝干趋升的灵动变幻，恰似火焰时高时低，忽左忽右地向上飘动。那是象征当今红红火火的年代 ，是激情燃烧的岁月，时代巨轮，滚滚向前，顺其者昌，逆其者亡。每个人都以崭新的姿态和拼搏的精神，不枉时代赋予的机遇，争分夺秒，争取最大成绩，将自己的一切做到最好。

此盆“对节白蜡”也可视为一把燃烧着的火炬。

火炬代表光明，冲出黑暗，照亮前程，勇往直前……

奥林匹克火炬代表世界和平，友谊，和谐相处，公平竞技。

抗日战争年代，在祖国大地，火炬代表我们万众一心，冒着敌人的炮火前进前进前进。

“二战”时期，物资匮乏，四处烽烟，生灵涂炭，世界一片灰暗，火炬的内涵有新的内容：打倒法西斯，解放全人类。当时的进步社团、学校不乏以火炬作图腾，铸入自家的标徽，鼓舞人心。此盆“对节白蜡”，所有赋予火炬的文化内涵和时代象征，它都齐了。

There are about ten trees whose roots of the trunk surround a giant rock and grow upward. One in the center is the highest to be backbone, while others are arranged by its sides in proper order according to the height, where the top forms is a triangular.

Each of the trunks is not upward in perfect straight as a bamboo tube, but in bend angle which is twirling and turning in its special space. The force is vigorous and powerful, just like there is a jet of invisible hot air rising. The branches indulge themselves to grow upward in a form of float with a high spirit of competition, a powerful uptrend, and an opened and expanded gaseous pulse. The trunk is rugged, natural and unrestrained. Where it stretches out, it seems to draw back, and where it draws back, it seems to stretches out. There are swells and scars on turning corners which expresses years of growth with the power of bone, flesh, tendon and pulse. The branch is plump, pliable and tough. With the intertwine of powerfulness and tenderness, line variety of light, heavy, slow and fast is appeared, which manifests a flourishing picture of lively and smooth rhythm with line floating upward dynamically.

The beauty of lines is a special aesthetic standard of Chinese nation: it considers that there is no specific meaning of lines while it is an artistic manifestation pattern with abundant connotation and infinite containment, which can be used for expressing feelings and meanings, and rich inner world. The beauty of lines has been wildly applied in the writing of Chinese characters, the creation of ink and wash painting and Penjing, etc. To feel the wonderfulness of power changes between light and heavy that contains in it, and generate sublimation of arts of beauty through rhythm changing between fast and slow. As for the beauty of lines, higher realm of illusion and mystery can only be achieved through feelings, which is impossible to fully express it with plain words. This is the artistic realm for appreciation of the beauty of lines which is ingenious beyond description.

This *Fraxinus hupehensis* is like the raging fire burning in a brazier that will never extinguish. Flexible change of the whole uptrend branches and trunks is just like the fluctuation of fire, which floats upward in swing. It stands for a booming age in today, which is a time of passion. Time won't stop. It will only roll forward. We shall follow the trend but not against it. Each of us shall appear a brand new attitude with the spirit of struggle. We shall not fail the opportunity of this age by chasing time and striving for greater achievements, and trying to be the best in all aspects.

This *Fraxinus hupehensis* can be also considered as a burning torch.

A torch represents light, which shall breakthrough darkness, illuminate future, and carry us forward...

The Olympic torch represents peace, friendship, harmony of the world and fairness of the competition.

In the age of Anti-Japanese War, here on our motherland, the torch represents that all the people of a single mind, which is that we shall face the artillery fire of enemies and move forward and upwards at all cost.

During the World War II, the material is deficient, the beacon fire is everywhere, the people are plunged into an abyss of misery, and the world is covered by darkness. The meaning of torch has its new content: overthrow the fascism, liberate whole mankind. Some of the advanced societies and universities at that time will consider the torch as their totems with signs of themselves embedded, thus to inspire people. This *Fraxinus hupehensis* contains all the cultural meanings and symbolizations of the age which are endowed to torch, with nothing left.

盆景素材的培育（十四）
——《盆景总论》（连载十六）
Penjing Materials Nurture
—Pandect of Penjing Serial XVI

培育好的盆景素材，首先要掌握什么是好的盆景素材。也就是说，必须要掌握盆景的美、构成要素、植物的生长原理等要素，方可培育出好的盆景素材。

文：【韩国】金世元 Author: [Korea] Kim Saewon

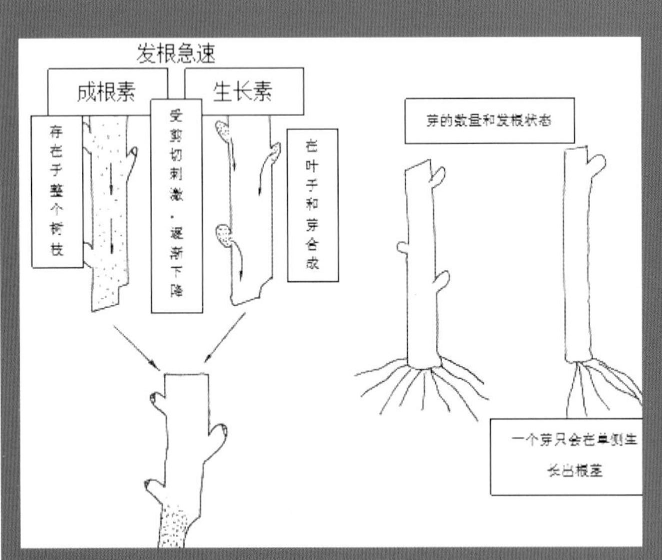

图 1 发根的原理

(3) 发根的原理：由于树种和扦插方法的不同，其效果存在较大的差异。如果要提高扦插效果，有必要先掌握发根的原理。

植物的发根过程需要植物激素、营养条件等各种要因。

根据温特 (Went:1938) 的假设，通过剪切刺激，树干或者树枝内影响生长和发根的物质——成根素(rhizocaline)会下降，同时，叶子和嫩芽内所含的植物激素——生长素（auxin）通过筛管部下降；在上述两种物质的共同作用下，在根部形成根原基（root primordium），促使根部的形成。

根据上述假设，通过实践我们可以获悉，选择长有一个嫩芽的插穗进行扦插，只有嫩芽的一侧长出根茎；同样，选择长有 2~3 个嫩芽的插穗进行扦插，根茎会均匀地长出。因此，盆栽素材用扦插繁殖要选用长有 2~3 个以上嫩芽的插穗。另外一个重要因素就是营养物质；成为植物生长活力源的碳水化合物或者用于根部生长的无机成分、氮化合物等都是必备因素。

挑选长势良好的树木，取健康树枝的中间部位作为插穗；由于营养和植物技术较为丰富，可有效地提高苗木的生长率。

① 植物激素：植物激素不仅会促进插穗的根茎发育，而且还会起到促进发芽、果蔬类增收、葡萄的无核化处理等作用。

Indolebutyric acid(IBA) 作为应用较为广泛的植物激素，不仅酵素的分解较少，而且可长时间保持稳定的使用效果。

目前在市面上销售的发根促进剂有萘乙酸（Rootone:α-naphthyl acetamide 0.4%）和吲哚丁酸（β-indolebutyric acid），1% 浓度适用于木本类，0.5% 浓度适用于草本类。

② 营养物质：插穗的发根到生长，糖等碳水化合物和氮化合物是必不可少的营养物质。

这些物质适量存在于插穗内；因此，用于盆栽素材生产的扦插无需采取其他特别措施。

在喷雾繁殖中，扦插嫩芽或者嫩枝时，碳水化合物和氮化物的含量比（C/N 率）会成为一定问题。

③ 阻碍物质：众所周知，植物体内不仅有生长促进物质，还有生长抑制物质。生长抑制物质是在 100℃温度下容易被破坏的酸性离子物质，基本类似于某种有机酸或者鞣酸（tannin）。

不同树种的生长抑制物质含量会有所差异；扦插效果不佳的栎树类、松树类等树种，其含量较高；相反，紫花槭、珍珠绣线菊、紫穗槐、水蜡树等树种，生长抑制物质的含量较低，插穗容易发根。

即使是相同的树种，根据树龄和树冠的部位，生长抑制物质的含量也会存在一定差异。

通常，随着树龄的增加，生长抑制物质的含量也会上升；另外，树冠上方日照条件较为良好的部位，其生长抑制物质含量高于中间部位。鞣酸与碳水化物的含量存在较大的相关性；容易发根的水菊、连翘等植物，鞣酸的含量较低，相反淀粉含量较高；发根率较低的红枫树、栎树等植物，鞣酸的含量较高，相反淀粉含量较低。

另外，汽油成分，松节油、香脂等特殊成分，树脂、氧化酵素等物质会加快受损处的氧化，而且还会阻碍发根。

④ 根原基的发生：根原基发生在根部生长出的部位；榉树、苹果树、中华绣线梅等树木在剪切之前，其树枝内已经发生根原基；与之相比，大部分植物在扦插后才会发生。

根原基逐渐与维管束相连成为根冠，并具备根部形态；接着，突破表皮，形成发根。

以木本植物为例，大多发生在形成层和筛部内，有些时候发生在愈合组织（callus tissue）。

根原基的发生受温度、湿度、光线等环境要素的影响，这些要素直接影响扦插效果；因此，扦插后直到发根为止，要维持插穗的活力。

【连载完】

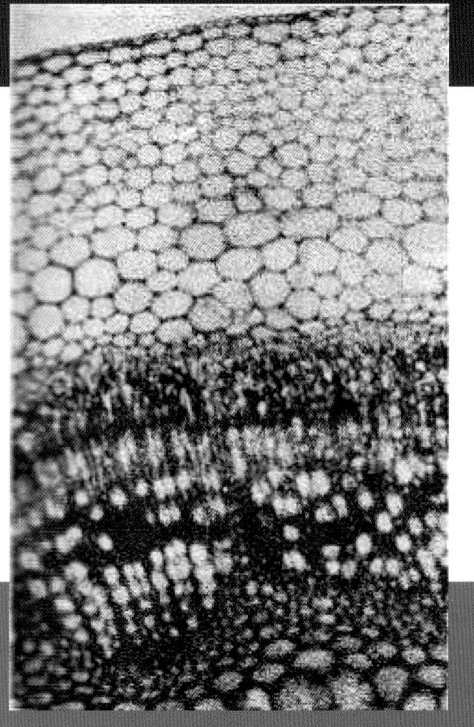
图 2 扦插前

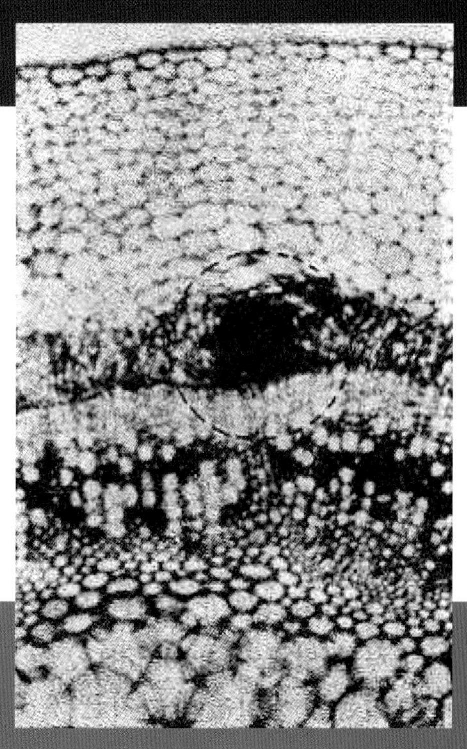
图 3 根原基的发生

图 1 主院落
Main courtyard

不是为了炫耀，而是为了人生

——从瓦茨拉夫·诺瓦克的私人院落里的一块草坪谈起

For Life not Showing off

—— Talking from a Piece of Lawn in the Private Courtyard of Vaclav·Novak

摄影: 苏放 撰文: CAT Photographer:Su Fang Author:CAT

图 2 矮墙内的庭院景观与墙外的自然美浑然一体，增加了庭院的纵深感
Courtyard landscape in the sunk fence and natural beauty outside form a unified entity, adding depth sense of the courtyard

这里是欧洲盆景大师瓦茨拉夫的家。一个位于捷克，很普通但是很美的小院落。

在现代人忙碌而功利的生活中，每个人都需要一个存放自己内心的领地，在欧洲人的人居方式里，这个领地就是草坪。

到过欧洲很多地方，你会发现这里的草坪特别绿，是绿得耀眼的那种，这种绿色里，一些深色的搭配使得这种耀眼并不会轻浮，而是让你的心变得更为沉静和内敛，每个地方都有这样的绿把你喧嚣的心一下子拉回到一个充满自然气息的静谧世界。相比中国大陆一些大气、豪迈，但处处粗陋、业余、漫不经心的私人盆景园的景观设计，欧洲人把并不昂贵的草坪看成是私人庭院的眼睛和心脏，

图 3 背面的一角仍是盆景管理台
The corner behind is still a Penjing management platform

图 5 从房间里看侧墙外
View outside the side wall from the room

图 4 草坪的出入口
Entrance and exit of the lawn

图 6 从房间里看主草坪
View the main lawn from the room

他们会为此投入最多的精力。不管是在皇家园林的鼻祖英国还是欧洲园艺的代表瑞士、法国……无论你走到哪里都会发现，草坪是欧洲庭院的主题和核心，尽管它在所有园林景观的组成中是最便宜的部分。这一点，和中国大陆的很多私人盆景园形成了鲜明的对比：大陆的盆景园动辄上百万的观赏石、名贵古树作装饰，锦鲤池里养着几十万元一条的锦鲤，水景水系混浊不堪，随处可见外露的黄土，粗陋而拙劣。

便宜，但是很完美；不喧嚣，但是很生活，这就是欧洲人的草坪。

有钱，但很俗气。看到这样的私人庭院景观的时候，你总是会很难过。

花了大钱的炫耀和竞赛并不是幸福。可惜，不少人现在还没领悟这一点。

在这个小院落里，可能没什么惊人的、值得炫耀的值钱东西，但是草坪很完美；高大的针叶树并不贵重，但突出了小院落里雍容优雅的气质；建筑不算豪

图 7 错落的植物搭配层次
Scattered plant collocation level

图 8 给盆景浇水是瓦茨拉夫每天要做的工作之一
Watering Penjing is one of Vaclav's work everyday

图 9 花丛
Flowering shrubs

This is the home of European Penjing Master Vaclav·Novak, a quite common but beautiful small courtyard in Czech.

In busy and utilitarian life of moderns, everyone needs a manor to store his inner heart which is the lawn in living mode of Europeans.

One will find most lawns in Europe are particularly dazzling green in which some brunet collocations makes it not frivolous and makes people feel quieter and internalized. Every place has such green putting one's blatant heart back to a tranquil world full of natural flavor. Compared with some landscape designs of individual Penjing garden which is magnificent, heroic but rash, unprofessional and casual, inexpensive lawns are regarded as eyes and hearts of individual courtyards by Europeans who will not spend too much effort. One will discover that lawns are the subject and core of European courtyards wherever he goes, Britain the founder of royal garden or Switzerland & France the representatives of European garden despite that lawns are the

图 10 建筑主面
The front of building

cheapest among the composition of garden landscape. This forms a bright comparison with many individual Penjing gardens in Chinese continent: continental Penjing gardens use stones and old trees which are of million Yuan for decoration; a koi worth thousands Yuan is raised in a koi pond; waterscape and water system are rude and clumsy muddy with exposed loess everywhere.

Cheap but quite perfect; quiet but with life style

图 11 精致小巧的盆景
Delicate and small Penjing

图 12 精致小巧的盆景
Delicate and small Penjing

图 13 精致小巧的盆景
Delicate and small Penjing

图 14 精致小巧的盆景
Delicate and small Penjing

图 15 盆景管理台
The Penjing management platform

华或令人感叹但是绝无任何粗俗之感；细节也许不很多，但是每个都做得很到位，绝不会给你那种看到哪里都是只做到一半的烂尾的感觉。

大门简洁朴素而又不失情趣。自然清新的气息随着满眼的绿色蔓延开来，大量的向阳性植物，深沉高耸的针叶乔木与亮丽低矮的阔叶小乔木混搭，使得这个院落和谐、宁静，又生趣盎然，爬满藤蔓的院墙很好地维护了院落的隐私，墙内的庭院景观与墙外的自然美浑然一体，增加了庭院的纵深感。主建筑的白墙红顶——强烈的色彩冲撞打破了满眼绿

图 16 商品盆景区一角
A corner of commercial Penjing area

图 17 台阶下的主草坪
Main lawn under footsteps

Cheap but quite perfect; quiet but with life style, European lawns are like this.

Rich but quite vulgar individual courtyard landscape would always make one feel uncomfortable.

Showing off and competing with much money but not happiness. Unfortunately, many people may not comprehend this at present.

There may not be some amazing valuable things that worth showing off in this small courtyard, but the lawn is quite perfect. Inexpensive tall coniferous trees highlight an elegant temperament in the courtyard; architectures do not contain any vulgar sense though they are not luxury or impressive; a few details are nicely designed and will not make one feel this is a hasty end.

The door is simple but with fun. Natural and fresh breathing spreads through eyeful green. Many turnsoles, deep & tall conifers and bright & low broadleaf trees mixing together makes a harmonious and tranquil courtyard with much fun. The courtyard wall which is full of cirrus well protects the privacy of the courtyard. Courtyard landscape inside the wall and natural beauty outside the wall form a unified entity which adds depth sense of the courtyard. The main building with white wall and red top—strong color collision breaks eyeful green vision inertia with changeful color level. There is a triangle scattering plant group in the left of the main building with the green and the red setting off each other; before the main building is a low platform connecting outside terrace and lawn in first floor of the residence with the platform spread by cobblestones and surrounded by thick crossties, scattering rough and original breathing.

图 18 瓦茨拉夫家的门口
The gate of Vaclav's home

图 19 外墙
Exterior wall

图 20 夏天纳凉水池一角
A corner of summer cooling pool

图 21 小盆景
Small Penjing

图 22 小盆景台
Small Penjing platform

图 23 修草坪是庭院日常生活中必不可少的环节
Mowing the lawn is indispensable in daily courtyard life

图 24 阳台
Balcony

在这样灿烂的阳光里，你只想在这里和自己最在乎的人在一起。

图 25 院落
Courtyard

图 26 院落
Courtyard

色的视觉惯性，色彩层次富于变化。主建筑左侧配置了一块三角形的植物群落，高低错落，红绿相映；主建筑前，一块低矮的台地衔接着住宅首层的室外露台和草坪，台地上铺满了卵石，并用粗大的枕木作围，散发着粗犷原始的气息。

建筑的右侧是草坪的一个出入口，路口相对狭小，隐蔽多荫，一棵高大的松树屏蔽了一部分视野，视线与脚步便自然被另一边一棵婀娜的红枫所转移，吸引人探秘寻幽，然而绕过红枫，却是草木葱茏，绿得发亮的大草坪。使得院落的空间分割在不大的地方有了变化和趣味。

院落的一角是瓦茨拉夫最钟爱的盆景区，干净整洁，井然有序，每一盆盆景都在这里得到了精心的照顾，每一棵都鲜活水灵，散发着纯粹的生命魅力。

主人在这里设置了吊床、秋千、儿童滑梯……坐在温暖和煦的阳光下，喝着咖啡看着儿童嬉戏，这让我想起了法国歌手 Claude FranGois 的一首经典老歌《Il Fait Beau, Il Fait Bon》（《晴空万里》），歌中唱道：Il fait beau, il fait bon. La vie coule comme une chanson…（晴空万里，生活像首流动的歌……）

在这样灿烂的阳光里，你只想在这里和自己最在乎的人在一起。

绿的耀眼的完美草坪、铺上了白桌布的餐桌，红酒、香槟、奶酪、一大盆色彩缤纷的绿色蔬菜和多彩的水果沙拉、美味的橄榄油醋汁、色味俱佳的主菜，阳光、亲友、宠物狗，还有北京似乎永远也不会再有的那种让人忍不住地深吸一大口的透人心脾的好空气……这是欧洲很多私人院落里最常见的一个场景。

喧嚣和竞赛并不是幸福，就像你有了钱却依然发现离幸福仍然很远。

幸福到底是什么？

一个人的院落往往会毫不留情地体现出一个人的品位和价值观。

有时，幸福只需要一块完美的草坪，并不需要花很多钱，你说是么？

图 27 院落
Courtyard

图 28 植物景观一角
A corner of plant landscape

In such brilliant sunshine, you merely would like to stay with those you care about most.

图 29 主草坪左侧建筑
The left building in the main lawn

图 30 主建筑前
The front of main building

There is an exit and entrance of the lawn in the right of the building. The road is relatively narrow and covert with many shades. A tall pine covers some view, thus vision and step are naturally transferring with a graceful red maple in the other side to discover secrecy. Bypassing the hill is luxuriant vegetation and a large green lawn. Those all make the division of courtyard space have change and fun in a small place.

A corner in this courtyard is Vaclav's favorite Penjing area which is clean and orderly. Receiving meticulous care, every pot of Penjing is fresh and alive and scatters purely living charm.

图 31 主建筑主墙
Main wall of mail building

图 32 主建筑左侧一角的三角形植物群
Triangle plant group in a corner of the left of main building

The owner set a hammock, a swing and a children's slide here...sitting under the warm sunshine and looking at children playing with a cup of coffee enables me to think of a classic song named Il Fait Beau, Il Fait Bon sung by a French singer named Claude François with the lyrics like this: Il fait beau, il fait bon. La vie coule comme une chanson...

In such brilliant sunshine, you merely would like to stay with those you care about most.

The dazzling green perfect lawn, a dining table with the white tablecloth, red wine, champagne, cheese, a large pan of colorful green vegetables & colorful fruit salad, delicious olive oil pickle, salad with good color and taste, sunshine, relatives & friends, pet dog and the fresh air which seems not to exist in Beijing forever and makes people refreshing after unbearably deeply breathing...this is most common scene in many individual courtyards.

Noisy and competition do not mean happiness, as if you find happiness is still far away from you though with money.

What is happiness on earth?

A courtyard of a person always reflects his taste and value without mercy.

Sometimes, happiness merely needs a piece of perfect lawn which does not need spending much money. What do you think?

图 33 主宅侧面
The side of main residence

图 34 走进院落的主道从里往过道外看
Look from inside to outside of the passageway after approaching to main stem of the courtyard.

紫砂古盆铭器鉴赏

Red Porcelain Ancient Pot Appreciation

文：申洪良 Author: Shen Hong Liang

图1 梨皮红泥蒲包口云足长方盆
长35cm 宽26cm 高14.5cm 申洪良藏品
Red-mud with Pear Skin Texture Calceolaria-shaped Rectangular Pot. Length: 35cm, Width: 26cm, Height:14.5cm. Collector: Shen Hongliar

梨皮红泥蒲包口云足长方盆。

"荆溪川石山人"的落款在古盆上很少见，但每次出现必伴随古盆中一等一的精品、铭品，已知存世不超过10个。

根据《江苏陶瓷工业志》记载：陈文伯、陈文居兄弟俩，清雍正至乾隆年间人，文伯号寄石山房，文居号荆溪川石山人，所制紫砂花盆畅销日本，久而不衰。据此推断"荆溪川石山人"就是清雍正、乾隆年前大名家陈文居。

此盆给人的第一印象是泥料、泥色和做工。红色的胎土、金黄色的煅泥颗粒以及高超的表面处理手法使得盆面给人以一种独特的梨皮肌理，也是目前已知紫砂器中最漂亮红泥梨皮。

图2 "荆溪川石山人" "Jing Xi Chuan Shi Shan Ren"

蒲包口本身是一种独特复杂工艺，在圆盆、腰圆盆上相对容易处理，在长方盆上可谓是挑战极限。与一般蒲包口盆相比，此盆蒲口略微向外飘出，盆主体肩部挺阔端庄，从上到下缓缓下收，膨胀张力十足。云脚和圈脚比例恰到好处，不仅当作盆的支撑点、面，同时使得盆的上下协调，盆整体的厚重、紧张感、张力感得到了加强。

盆底双层底的工艺和开孔方式也是典型的雍正、乾隆年间做法。落款"荆溪川石山人"，端庄工整精细，略懂篆刻知识的藏家一眼就能分辨出此章必出于名家之手。

在清康熙、雍正、乾隆鼎盛时期，爱好盆景的文化阶层参与了盆的创作，此盆感受到得那份内敛的美，充分体现了文人气息。在已知盆中，此盆是唯一的"荆溪川石山人"红泥梨皮蒲包口盆。

日本盆景界大家高木在20世纪90年代曾出价1000万日元求购此盆，但最后没有如愿。此盆是目前已知中型盆中，做工最精细、泥色最漂亮的名家盆之一，足以让国人感到自豪。更让人高兴的是，经过几百年的海外漂泊，此盆已在今年悄然回归故里。

成为中国盆景艺术家协会的会员，免费得到《中国盆景赏石》

告诉你一个得到《中国盆景赏石》的捷径——如果你是中国盆景艺术家第五届理事会的会员，从下一个月起我们开始赠送给您。

成为会员的入会方法如下：

1. 填一个入会申请表（见本页）连同 3 张 1 寸证件照片，把它寄到：北京朝阳区建外 SOHO 西区 16 号楼 1615 室中国盆景艺术家协会秘书处（一定要注明“入会申请”）邮编 100022。
2. 把会费（会员的会费标准为：每年 260 元，港澳台除外。）和每年的挂号邮费（大陆每本 6.3 元，每年 12 本共 76 元；港澳台每本 50 元，每年 12 本共 600 元）汇至中国盆景艺术家协会银行账号（见下面）。
3. 然后打电话到北京中国盆景艺术家协会秘书处口头办理一下会员的注册登记：电话是 010-5869 0358。

会费收款银行信息：
开户户名：中国盆景艺术家协会 开户银行：北京银行魏公村支行 账号：200120111017572
邮政地址：北京市朝阳区建外 SOHO 西区 16 号楼 1615 室 邮编：100022

中国盆景艺术家协会会员申请入会登记表　　证号（秘书处填写）：

姓名	性别	出生年月	照片（1 寸照片）
民族	党派	文化程度	
工作单位及职务			
身份证号码	电话	手机	
通讯地址、邮编		电子邮件信箱（最好是 QQ）	
社团及企业任职			
盆景艺术经历及创作成绩			
推荐人（签名盖章）			
理事会或秘书处备案意见（由秘书处填写）：　　年　　月　　日			

备注：请将此表填好后，背面贴身份证复印件连同 3 张 1 寸照片邮寄到北京市朝阳区建外 SOHO 16 号楼 1615 室 邮编 100022。
电话：010-58690358，传真：010 - 58693878，E-mail: penjingchina@yahoo.com.cn。

《中国盆景赏石》——购书征订专线：（010）58690358

订阅者如何得到《中国盆景赏石》？

1. 填好订阅者登记表（见附赠的本页），把它寄到：北京朝阳区建外 SOHO 西区 16 号楼 1615 室中国盆景艺术家协会秘书处订阅代办处，邮编 100022。
2. 把书费（每年 576 元）和每年的挂号邮费（大陆每本 6.3 元，每年 12 本共 76 元；港澳台每本 50 元，每年 12 本共 600 元）汇至中国盆景艺术家协会银行账号（见下面）。
3. 然后打电话到北京中国盆景艺术家协会秘书处《中国盆景赏石》代订登记处口头核实办理一下订阅者的订单注册登记，电话是 010-5869 0358 然后…… 你就可以等着每月邮递员把《中国盆景赏石》给你送上门喽。

中国盆景艺术家协会银行账号信息： 开户户名：中国盆景艺术家协会 开户银行：北京银行魏公村支行
账号：2001201110175 72

《中国盆景赏石》订阅登记表

姓名：________ 性别：____ 职位：________

生日：____ 年 ____ 月 ____ 日

公司名称：________

收件地址：________

联系电话：________

手机：________ 传真：________

E-mail（最好是 QQ）：________

开具发票抬头名称：________

汇款时请在书费外另外加上邮局挂号邮寄费：大陆每本 6.3 元，每年 12 本共 76 元；港澳台每本 50 元，每年 12 本共 600 元（由于平寄很容易丢失，我们建议你只选用挂号邮寄）。

书费如下：每本 48 元。

☐ 半年（六期） ☐ 288 元 ☐ 38 元 ☐ 300 元

☐ 一年（十二期） ☐ 576 元 ☐ 76 元 ☐ 600 元

您愿意参加下列哪种类型的活动：

☐ 展览 ☐ 学术活动 ☐ 盆景造型培训班 ☐ 国内旅游（会员活动） ☐ 读者俱乐部大会

☐ 国际 旅游（读者俱乐部活动）

讣 告

中国著名盆景艺术家、书画家、盆景艺术理论家、实践家、教育家、湖北盆景的开创者、中国动势盆景的一代宗师、中国盆景艺术大师贺淦荪先生，因病经医治无效，于2013年9月6日下午3时10分在武汉与世长辞，享年89岁；贺淦荪先生为中国盆景事业的继承和发展做出了卓越贡献，他的逝世是中国盆景事业的巨大损失：贺淦荪先生追悼会将于2013年9月10日上午9时在汉口殡仪馆举行；特此讣告！

贺淦荪大师治丧委员会

2013年9月7日

柯家花园仿古石盆 系列欣赏

The Appreciation of the Ke Chengkun's Antique Pot Series

1-35# 仿古石盆　长1.93m 宽1.15m 高0.88m
榕树 高 2m 宽 2.3m 直径 0.5m 柯成昆藏品 柯博达摄影
1-35# Imitation Ancient Rock Pot. Length: 1.93m, Width: 1.15m , Height: 0.88m.
Ficus. Height: 2m, Width: 2.3m, Diameter: 0.5m.Collector: Ke Chengkun.
Photographer: Ke Boda

本年度本栏目协办人：李正银，魏积泉

赏石中国

CHINA SCHOLAR'S ROCKS

“天险” 九龙璧 长38cm 宽25cm 高45cm 魏积泉藏品

“Natural Barrier”. Nine Dragon Jade. Length: 38cm, Width: 25cm, Height: 45cm. Collector: Wei Jiquan

“貔貅” 大化彩玉石 长33cm 宽23cm 高13cm 李正银藏品 苏放摄影

“Pixiu”. Macrofossil. Length:33cm, Width: 23cm, Height: 13cm. Collector: Li Zhengyin, Photographer: Su Fang

“荡中寻宝” 广西天峨石 长15cm 宽11cm 高23cm 魏积泉藏品
“Treasure Hunt in the Swing”. Tian'e Stone, Guangxi. Length: 15cm, Width: 11cm, Height: 23cm.
Collector: Wei Jiquan

新疆巴里坤风光

中国古今名石简谱（连载七）

Chinese Famous Rocks Notation (Serial VII)

文:文甡 Author: Wen Shen

"心里美"哈密石友的小品石

笔者在哈密魔鬼城戈壁滩寻石

和田卡瓦石 小品

"瓜"玛瑙 小品

和田青玉籽 长15cm

新疆和田墨玉籽 长10cm

"富（父）"和田羊脂玉籽

"台"长12cm

哈密南湖大漠石戈壁滩

（二）哈密大漠石

哈密是新疆的东大门，古丝绸之路过玉门、出敦煌，便来到被称为"西域禁喉"的新疆大漠石主产地哈密市。哈密大漠石产区北至伊吾，西至鄯善，南至大黑山，在这广阔的大漠戈壁中，蕴藏着丰富的奇石资源，成为中国大漠戈壁名石的宝库。

1.哈密大漠石的开发与现状

哈密奇石的市场，始于20世纪末期，标志着哈密奇石收藏己经初具规模。此前不断有外地石商到哈密，以每千克几角钱的价格，成车成吨地将大漠奇石运走。哈密的石友、石农在这奇特的交易中，逐渐明白了这些大漠奇石的价值，于是大规模的奇石开发展开了。

在哈密南湖戈壁滩，在东天山伊吾县淖毛湖，在众多产奇石的大漠中，哈密人以无畏的精神，纵深大漠几十、数百千米，无数次进出大漠，将这些亿万年前的精灵，运出戈壁，创造了哈密奇石令人瞩目的成就。

哈密大漠奇石十年来的开发，发现了众多的石种，开拓了市场，举办了石展，将哈密奇石带到各地，推向了全国。在取得这些成就的同时，奇石资源也日益减少，供求矛盾越发突出，人们意识有所突破，藏精品、玩文化的风气形成，哈密人对大漠奇石的鉴赏水平踏上一个新高度。

2.哈密大漠石的品种与鉴赏

哈密的大漠戈壁石品种非常丰富，散布在广袤的大戈壁中，但每类石种的总量却不大，造成名品的稀缺。同时新的石种又不断出现，又给石界带来惊喜。于是哈密就成为石友淘宝的乐土。多年来，哈密大漠中最为人津津乐道的，有木化石、风凌石、玛瑙、泥石、蛋白石等品种。

(1)木化石

哈密木化石，主要产于哈密以南大约100km南湖煤矿周围，是最优质的硅化木。做为有较高赏玩价值的哈密硅化木老货，具备以下的特质：

其一是地表石。裸露在大漠中的硅化木，经过风沙长期磨砺，表皮光泽润滑，与地下挖出并经过加工的的木化石，不可同日而语。

其二是硅化程度高，造型、纹理变化多样。

其三是色彩有棕、黄、红、黑等色谐调柔和，有玛瑙质地的通透光泽。

(2)风凌石

哈密风凌石，主要产于哈密以南大约300km的马蹄山一带。哈密南湖多处都有发现，是哈密奇石中产量较多的品种，主要有以下特点：

其一是造型丰富，变化多样。哈密风凌石有景观、古堡、人物、动物、器物等各种形态，细微处结构巧妙，极具观赏、收藏价值。

其二是色彩多样，有黑、灰、白、红、黄、褐等颜色，更兼有俏色、多色搭配，常有出其不意的视觉冲击。

其三是质地优秀。哈密风凌石质坚而润，部分风凌石呈半透明状，大部分都有天然石皮。

近年来又发现玉质风凌新品种，除具有风凌石的特质外，

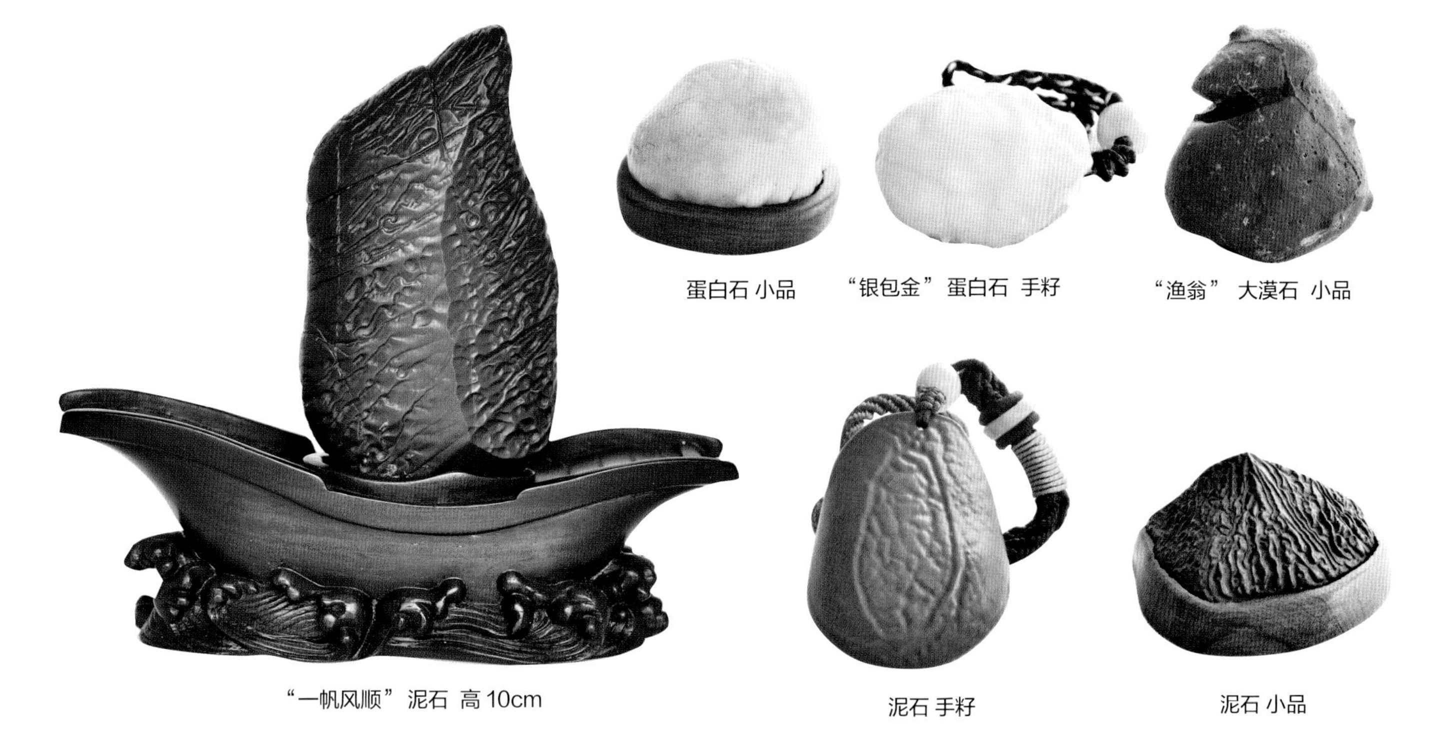

蛋白石 小品

"银包金"蛋白石 手籽

"渔翁" 大漠石 小品

"一帆风顺" 泥石 高 10cm

泥石 手籽

泥石 小品

"洞天" 大漠石 小品

泥石 小品

蛋白石 高10cm

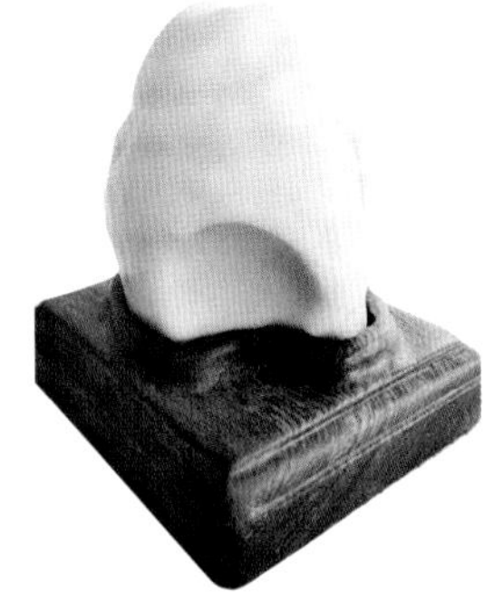
蛋白石 小品

更加鲜亮光泽、通透娇美，受到石界的喜爱。

(3)玛瑙石

哈密玛瑙石，主要产于哈密以北，东天山伊吾县的淖毛湖一带，距哈密260km，翻山越岭，道路十分艰难。淖毛湖玛瑙滩面积约40km²，玛瑙质优色艳，突出的特色有以下几方面：

其一是品种众多。有缠丝玛瑙，平行柔美的曲线，构成各种图案，非常别致；有碧玉玛瑙，是碧玉与玛瑙的共生体，色纹相映衬，富丽雅致；有闪光玛瑙，转动时可见纹理间光带闪动，富有情趣。

其二是色彩丰富。哈密玛瑙色彩娇艳，有红、黄、绿、黑、白等各种色彩齐全，更有俏色如夹心糖果、充馅蛋糕，让人垂涎欲滴，成为追捧的热门玛瑙石种。

其三是纹理变幻无穷。哈密玛瑙石的纹理变化多样，有缠丝纹、水浪纹、彩带纹、圈眼纹、云雾纹等不一而足，形成哈密玛瑙的特色。

(4)泥石

哈密泥石，是各种泥石中的最优质的熟泥石，质、色如紫砂。哈密泥石产于哈密以南大约90km的南湖戈壁滩上，目前发现的3个泥石坑，每个面积约20km²。南湖泥石有以下几个特征：

其一是质地细腻，观感上佳。

其二是颜色以紫棕、绿色为主，色泽干净无杂色。

其三是纹理流畅，富于变化。

其四是石皮明显，易出包浆，手感极好。

最近在距哈密以南400km的戈壁深处，又发现了一处新泥石产地。这种新泥石以纯黑和铁灰色为主，纹理更加深刻，包浆也更加古旧。

哈密泥石以它似古旧紫砂的老到，古朴醇厚的味道，成为文房中的清供。

(5)蛋白石

哈密蛋白石，早在10年前，就在哈密以南数百公里的大黑山发现，近年在哈密以南的戈壁腹地又有发现。这种白色而带漆色的大漠奇石，有着独有的特色：

其一是洁白如玉似瓷。哈密蛋白石，有的半透明温润如玉有蜡质光泽，有的洁白如瓷有玻璃光泽。另有一种黄如鸡油，极为鲜亮。也有银包金相融一体，更为奇特。

其二是哈密蛋白石都有沙漠漆沁色，多为黑、红色，与白色质地相映成趣。

其三是哈密蛋白石有老道的石皮，大多可做天然手件把玩，手感如玉，古味很足。

3.哈密大漠石的收藏与价值

哈密是新疆大漠奇石的最大产地，由于开发较晚，也是中国当代赏石的后期宝库。

(1)大漠石的收藏

哈密风凌石与其他戈壁石比较，变化更加多样，颜色更加丰富。不足之处是部分风凌石棱角分明，可观赏而不耐盘玩。后来发现的玉质风凌石，以其温润的特质，弥补了这些许缺憾。

哈密大漠的独特奇石，是泥石与蛋白石。泥石质、色如紫砂，细腻且富于纹理。置于案头古朴典雅，握于手中包浆醇厚。

哈密南湖大漠石戈壁滩

大漠石 长8cm

风凌石 长12cm

泥石 小品

蛋白石有似和田老玉者，通透朦胧，温润而有柔和的蜡质光泽，并且多有如和田籽玉的沁皮，更添几分韵味。于手中盘玩，古气盎然。哈密大漠泥石与蛋白石，古朴醇厚的味道，适合把玩的特质，与中国传统文玩十分契合，是文房清玩的首选。

哈密东天山北麓淖毛湖一带的玛瑙石，多有如糖果、糕点等美食的俏色玛瑙，色彩娇嫩，令人垂涎，是一种极有特色的石种，成为有心人的专宠。

(2)大漠石的价值

哈密大漠石的开发已有十几年，价格从刚开始的整车千八百元，到现在每件几百、几千、几万元。2008年，一件20cm左右的玉质风凌石，以40多万元的价格成交，这是迄今为止哈密大漠石成交的最高价。

哈密大漠石开发相对较晚，其特点是地域广阔、品种多、个种产量不大、不断有新石种发现。哈密大漠石目前仍有存量，价位不高，市场升值潜力巨大，被称为赏石的后期宝库。

【未完待续】

“钓鱼台”大漠石 小品

“元宝” 大漠石 小品

蛋白石 小品

大漠石 小品

图纹石的质地之美解析

Analysis of the Beauty of the Texture of The Figure Proluta

文：雷敬敷 Author：Lei Jingfu

质地，也叫石质，在含义上有广义与狭义之分。广义的质地是一个复合的概念，它包括奇石内在品质的硬度、细度、韧性、透明度及外在表现的润度、光泽、风化膜、风化痕等。狭义的质地专指细度。本文指广义的质地。

图纹石的质地之美，既在于质地本身所具有的相对独立的审美属性，也在于质地对纹理与色彩的影响。我们从质地组成因素的审美特性，质地对图纹石石画影响的对比分析和图纹石的石肤与肌理之美这三个方面来进行说明。

一、质地的组成因素及审美特性

1. 硬度

硬度反映石质的坚硬程度，通常以摩氏硬度来衡量。摩氏硬度是以不同硬度的矿物为相对的硬度标准，将岩石的硬度分为十级，构成摩氏硬度计。

奇石硬度通常认为以摩氏硬度4°~7°为宜，对应标准矿物在萤石至石英之间。硬度过高不易在风化过程中变得圆润，过低则易风化磨损，也不便于保藏。

奇石的硬度以及那亿万年的造岩与风化中历练的沧桑感，在审美上一直是作为品格坚强和历久弥新的精神境界的一种象征。

图1 "静夜思" 雷敬敷藏品

2. 细度

细度是组成岩石的颗粒在大小上的分布情况，质地越细，就是细颗粒的比例越大。我们可以将奇石的细度分为3级。其中肉眼可分辨者为粗；10倍放大镜才可以分辨者为细；10倍放大镜也不能分辨者为细腻。通常砂岩为粗，泥岩为细，玛瑙质地是一种胶态物，为细腻。细腻的质地一般受人喜爱，常以少女凝脂一样的肌肤作为比喻。

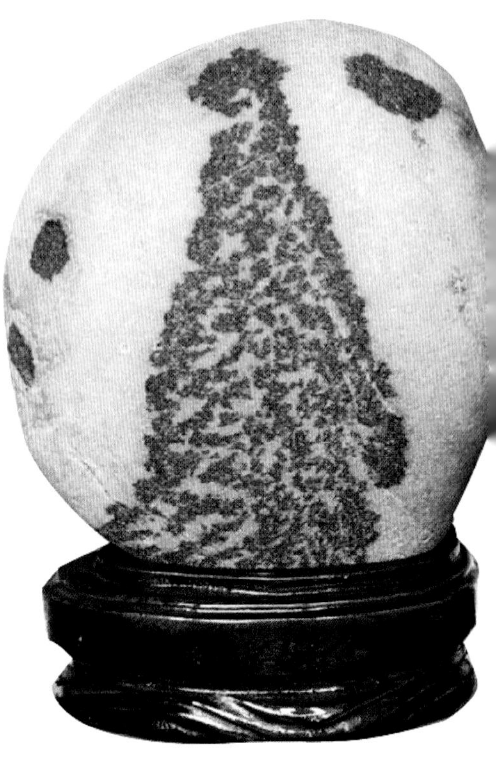

图2 "穿裘皮的少女" 雷敬敷藏品

3. 润度

润度其实是有较高硬度和细度的奇石因其光滑的石面上的天然包浆所表现出来的一种滑润感。所以粗糙的或完全无包浆的石面无润可言。通常硬度较高、结构致密、水洗度好的奇石润度高，如果又有一定的透光性，便有玉的温润感了，通常视为上品。润度可以分为不润、润、温润3级。

4. 光泽

光泽是石面在外来光线照射下所产生的光斑的状况，其强弱取决于对外来光线的反射率、吸收率和吸收系数。对光泽的判别方法是将奇石的主观赏面对着阳光看有无反射的光斑及光斑的强弱情况，它与地学上以矿物或岩石的新鲜断面上光的反射情况来定义光泽不同。方伟将奇石的光泽分为玻面光泽、釉面光泽、蜡面光泽、陶面光泽和土面光泽5级。以卵石为主的图纹石多为陶面光泽，在阳光下形成的光斑柔和并有渐变光晕；玛瑙质和玉质的图纹石多为蜡面光泽。

不同的光泽会给我们不同的审美感受。玻面光泽如镜面反光，张扬，璀璨，仅见于个别的戈壁玛瑙。釉面光泽有模糊的镜面效应，光洁、富丽，一般见于石质坚硬致密的不透明石种，如大化石、彩陶石、沙漠漆石。蜡面光泽是半透明或透明石体反射、透射及散射出的光泽，是玉质类奇石所特有，常见于蜡石、戈壁玛瑙。陶面光泽儒雅、醇厚，常见于河滩卵石。土面光泽是石面几乎无反光的光斑，光线于其表面几乎完全是散射，显得质朴、粗犷。

5. 韧性

韧性是岩石抵抗外力破坏的能力。韧性强的岩石不易碰伤、碎裂。若硬度高，又韧性强，可谓有坚韧的品格。新疆和田玉便是其中的佼佼者，硬度6°~7°，韧性又特别强。和田玉观赏玉多是玉质的图纹石，由于韧性好，边缘少有断口，石面上也少有影响观赏的裂隙纹之类的天残，因而弥足珍贵。

6. 透明度

透明度是石体透过光线的能力。透明度分透明、半透明、不透明3级。测定方法是将微型手电筒的光照面贴紧石面后，观察手电筒边缘透出的光晕的宽度。光晕大于1cm者为透明，小于1cm者为半透明，无光晕者为不透明。

从审美的角度看，透明度是“奇”的珍稀性的重要方面，特别是对于一般的河滩卵石，透明度高是“玉化”程度高的一个指标。

图3 “东港余辉” 赵卫东藏品

图4 “塔里木河之秋” 唐传林藏品

7. 风化膜

风化膜又叫风化漆，是外力的化学风化作用在石体表面所形成的氧化物的薄膜层，也称之为天然包浆。据方伟研究，风化漆有3种类型：一是釉彩漆，色彩绚丽斑斓、光泽明亮华贵。由于形成环境独特，仅见于广西大化石，内蒙古荒漠漆石；二是雾彩漆，是一透明或半透明的如纱似雾的带色薄膜，如部分河滩卵石；三是透明漆，为透明状薄膜，多见于河滩卵石和新疆风棱石，风化膜是图纹石天然性的标准之一。

8. 风化痕

风化痕也叫风化面，它是散布于整个石画上的细微的风化痕迹。奇石不论如何坚韧、圆滑、细润，若仔细观察，天然的石肤上通常都会有细微的凹坑麻麻点点的遍布着，俗称“麻坑面”或“皱皮面”，就像人体皮肤上的细纹与毛孔。长江石中硬度高的玉化了的硅质岩，石画上还有指甲痕状的“指甲纹”，有的石友称之为“指甲玉”。

风化痕与风化膜一样，也是图纹石天然性的特征之一。由于它比风化膜更易对比判断，是考察图纹石是否有人为加工的最重要的标志。

二、质地对图纹石石画影响的对比分析

“有比较才有价值”，质地对图纹所成的石画的影响我们从粗与细，润与糙，透与实，光与哑这4个方面对比分析。

1. 粗与细

一般而言，质地细有利于色彩的表现和对比度的提高；相反，质地粗时，因反射光中有相当多的漫反射，使色彩晦暗，边界模糊。

其实，对于图纹石而言，质地粗细皆与石画配合得当为好。我们以笔者所藏的两枚长江石来作对比分析。

一枚质地细，属瓷质石肤，在“静夜思”的石画上一老者低眉凝目，美髯飘然，深沉的夜色中人，在细腻质地映衬下于醇厚之中又显得清雅（图1）。

另一枚质地粗，属砂质石肤，石画上身着裘皮大衣的少女，蓬松的发际，修长的身材在沙质的粗糙背景下，显得古拙纯朴而又迷离朦胧，正合石画主人公深闺冬来春意多的意蕴（图2）。

作为图纹石之一的雨花石历来有粗石、细石之分。过去人们对玛瑙质的细石倍加珍爱，而对普通卵石状的粗石不屑一顾。近年来，随着审美意识的变化，人们逐渐发现了粗石之美，认识到粗细之间其实各有所妙。著名雨花石

收藏家刘水认为，细者如江南丝竹，粗者如京韵大鼓，有风格之别，无高下之分。

2. 润与燥

润是玉化的石肤上的天然包浆在有一定反光但又不太强烈的情况下所表现的一种滋润感，这其实是由视觉引发的触觉的一种通感的现象，就好比熬糖时的气息使我们感觉到甜味一样。

额河石"东港余辉"（图3），是润的实例。傍晚，在金色的夕照下，港口的大型机械已成一片灰暗的剪影，矗立在空中，倒映在水里。晚霞之上是余辉映照的天空，晴朗而又滋润。

在审美时滋润感会拉近人石的距离，会使人产生触摸的冲动。在润的情况下图纹的形象与色彩会更加明朗、清晰。这时赏石者常会情不自禁地赞叹："多好的皮呀！"

相反，干燥的石肤却给人以排斥感。就图纹石而言，干燥的石肤多半是因酸洗而破坏了天然包浆所致。这时就有所谓的养石之说了。养什么？养护出天然包浆式的人工包浆，诸如手养、油养等，反过来也说明了"润"之重要性。

3. 透与实

透明与半透明的质地不但体现了玉或类玉石的珍稀特性，而且由于反射光与透射光的共同作用，会使色彩纹更加绚丽。半透明的玛瑙质的图纹石在水中赏析，色彩会显得光怪陆离，晶莹剔透；如果要在空气中也有较好的视觉效果，那就要有很好的水洗度；光滑而且平顺。图4的和田观赏玉"塔里木河之秋"，在青白玉的背景上水镀的"糖色"变幻成一片金黄、橙黄和棕褐交映的秋景，在半透明的温润氛围下，这秋景在热烈之中又趋于婉约与宁静。

完全不透明的实，是绝大多数河滩卵石的属性，也自有它的优势所在，那便是实在与充实的品格。图5河滩卵石"山村人家"实而不透，白色的石画上，黑色素描出近处的坡地蜿蜒起伏，远处的树丛中露出了几栋黑瓦白墙的农舍，整个石画显出朴实无华的敦厚。

图5 "山村人家" 俞兆虎藏品

4. 光与亚

光泽反映出石质的硬度与细度。方伟认为，抗风化力强的硬度高的质地，在自然流体介质的风沙或水沙作用下表现为抛光的作用，越磨越光滑。而对于软质的石质却因磨蚀而形成粗糙的磨砂面，造成光的漫反射而降低光泽度。

其实对于图纹石而言，过度的光泽是不利于石画赏析的，强烈的反射光斑会影响石画的整体感。图6的长江蜡石是在玉化的石质上，以紫红的天际，层层的白云构成一幅高雅的石画。笔者有诗赞曰："千秋忠义薄云天，一石浩然顾盼间。质比蓝田应为玉，纹如霞蔚疑是仙。"但由于该石反光面太大，给赏析带来些许遗憾。

图7的额河石"板桥遗墨"虽然同样石质坚硬、细腻，但因其石皮上遍布的麻坑面的漫反射作用，减弱了反射光的强度而具有亚光的性质，暗色背景下浅绿色的竖立的干、横斜的枝，交错的叶构成了一个几无反光干扰的完整的赏析——面。

所以，对于硬度高、水洗度又好的图纹石不宜以油脂护面，以免造成强烈的光斑。目前形成亚光的效果还是以涂蛋清为宜。

三、图纹石的石肤与肌理之美

图纹石的石肤与肌理有其特殊性，我们分别就石肤的分类及其风格气质，肌理的意义与笔墨意趣加以阐述。

1. 石肤的分类及其风格与气质

石肤也叫石皮，是质地诸因素在石体表面上的综合体现，是我们对质地可观赏、可触摸的具体的感性的外表形象，是质地诸因素综合审美的表征。

我们可以对以河滩卵石为主的图纹石的石肤作以下大致的区分，它们有各自不同的风格与气质。

（1）玉质石肤：石质坚硬、细腻、温润，光泽较强，蜡面光泽，透明或半透明，气质高贵、典雅。

（2）类玉质（玉化）石肤：石质坚硬，细腻滑爽，具蜡面和陶面的光泽，微透明（亚光）至不透明。气质富丽，高雅。

（3）瓷质石肤：石质较硬，结构颗粒细，肉眼不可分辨，手感较顺滑，光泽较弱，为陶面光泽，不透明。气质醇厚，清雅。

（4）沙质石肤：石质较软，结构较粗，颗粒肉眼可辨，不透明，手感涩滞，土面光泽无光斑。气质纯朴，古拙。

由于不同的石肤有不一样的视觉和触觉的感受，所以同一石类可以按石肤作进一步的细分。如长江石中一类红色的长江卵石，对于玉质或类玉质的长江石石友常称之为长江红碧玉，而于沙质石肤的则称之为麻岗红。

2. 肌理的意义与笔墨意趣

如果说图纹石的形的外廓好比一幅画的形状与尺寸的话，那么图纹石的石肤就是这绘画的基质材料的表面状况。同样一幅绘画作品在纸面上与布面上的感觉是不一样的。吴凡的著名木刻作品“蒲公英”，就是因为拓印在宣纸上，才表现出小女孩红润的脸庞和被吹散的蒲公英的飘逸之态。韩美林将国画的技法用在道林纸上，他画的动物形象较之于在宣纸上另有一番情趣，形成了自己独特的风格。

雕塑家说的肌理是雕塑材料的质感，绘画家说的肌理是不同的绘画材料在笔墨颜料的绘涂下所表现的笔触，我们这里所说的肌理是图纹石的石肤与纹理色彩共同作用下所表现的一种视觉效应，类比于绘画的笔触，表现的是笔墨意趣之美。

我们前面谈到的石肤的分类及风格气质，是就石肤本身而言；这里谈到的肌理是不同的石肤在各种性质的纹理色彩的“描绘”下的“笔墨”表现。

例如，图4的点涂线描，与晕染交替，用笔娴熟灵动；图6则在紫红的背景色尚未干透的情况下，又醮浓浓的白色颜料一挥而就，红白边缘自然交融，狂放之中又显柔和之态；图1犹如精细纸张上所绘的铅笔素描，用笔精致细微；

图6 “义薄云天” 宋旭东藏品

图2则是用稀薄颜料在粗糙表面上的快速写意，令粗糙的画布质感带来一种雾里看花的梦境般的迷蒙。

又如图7的“板桥遗墨”，以大刀阔斧般的浓墨挥洒，酣畅淋漓。图5的“山村人家”却是焦墨皴涂，其用笔先后，历历可辨。图3的“东港余辉”则是西洋画风，用笔遒劲，笔触尽显。

一是石画边缘的断口破损，若天残之后又经过一定时间的水磨沙砺，破裂断口已有一定磨圆度，成为了“老残”、“老崩”，而且破损面不大，通常是可以接受的。二是石画的裂隙破损，有时还伴有风化过程的沉积物。对这类天残的具体分析，一看这裂隙破损与整体图纹的关系，二看赏析者能否从中给一个恰当的说法。如果能，那就是化腐朽为神奇了。如果不能，那就只能以“人无完人，石无完石”来自我安慰。

至于人为的破损，那当然都是不可取的，留下的是更多的遗憾。

更正

《中国盆景赏石·2013-8》52页《树木盆景的创作应尊重树种个性特征》一文中，53页左上图为“榔榆　胡平春藏品”，右下图应为“三角枫　胡平春藏品”。103页《独坐黄昏》一文配图藏石者为彭祖国。因本书校对疏忽，特此更正，并向作者和广大读者致歉！

图7 “板桥遗墨” 陈如意藏品

中国鼎2013

中国（古镇）盆景国家大展

通告

关于 2013 中国盆景年度之夜的介绍

2013年9月29日晚上的中山古镇，中国盆景的一次全明星式的年度盛会将隆重亮相！中国盆景艺术家协会不断发展壮大的会长团队将全体登台亮相，超级盆景的著名收藏名家们、中国最专业的盆景制作人们、中国各地的盆景界重量级人物将群星闪烁，照耀中国。届时，中山古镇将成为2013年中国盆景年的“中国焦点”之一。

“2013中国盆景年度之夜”将是一个中国盆景的年度盛典。它将向全中国推出中山古镇的盆景城市名片，并将推动下一步在中山古镇搭建大型盆景产业化项目平台的发展计划。这一盛典将邀请众多对中国盆景艺术、中国盆景协会组织、盆景制作和盆景市场经济做出了贡献或有特别意义的重量级的嘉宾参加。晚会上将进行“2013中国鼎”、“2013中国盆景年度先生”、“2013中国盆景年度城镇”、“中国盆景名镇”、“2013中国盆景年度协会、中国盆景艺术家协会年度会员俱乐部、中国盆景艺术家协会年度会员小组”等国家盆景称号和大奖的颁发仪式，届时，所有国内外嘉宾出席者都应着符合国际晚会习俗的黑色晚会正装或晚装（中山服正装、西装、晚装均可），以最隆重的氛围来分享这次年度盛典。

中国盆景艺术家协会

2013年6月

中国盆景艺术家协会全体会员及全国所有盆景爱好者们，为打造面向全球的中国盆景国家展的标志性品牌，向全世界展示中国盆景的国家形象，2013 年9月29日至10月3日，中国盆景历史上的首次“中国（古镇）盆景国家大展”将在广东省中山市古镇镇隆重推出，多项重大活动项目也将一并推出，详情如下：

关于2013中国（古镇）盆景国家大展的介绍

国家大展的名称释义与定位：“中国（古镇）盆景国家大展”是由中国盆景艺术家协会(CPAA)、中山市人民政府在2013年开始推出的国家盆景展览品牌，定位（办展目标）是：国家一级协会主办的代表中国盆景最高专业水准的国家展。

主办单位： 广东省中山市人民政府 中国盆景艺术家协会
承办单位： 中山市古镇镇人民政府 中山市农业局
中山市林业局 中山市海洋与渔业局 广东省盆景协会
协办单位： 中山市南方绿博园有限公司
支持单位： 中山市盆景协会、中山市古镇盆景协会及中国盆景艺术家协会在全国各省市的合作协会
地点： 广东中山市古镇镇灯都古镇会议展览中心
时间： 2013 年 9 月 29 日至 10 月 3 日
展会日程：
9 月 26~27 日：布展
9 月 28 日：评奖、嘉宾报到
9 月 29 日：开幕式及举办单位、会员、嘉宾展场参观
9 月 29 日晚 6 时 30 分：中国盆景年度之夜
9 月 30 日：对公众开放展场，嘉宾上午参观展览和制作表演，下午继续参观或举办相应的学术交流活动
10 月 1~3 日：展览继续开放
10 月 4 日：撤展
参展作品规格及选送：

国家大展的参展作品规格尺寸标准为：高1.5m内（文人树最高可放到1.6m以内，文人树的评判标准由评委组投票决定），盆长1.8m内，悬崖飘长1.5m内。

送展的作品均要配以相衬尺寸的几架（一般情况下10~20cm为宜），盆景的展前修饰、美化、配盆和几架的搭配都将计入该盆盆景的参赛评分中。其中，任何没有做过精心的展前修饰工作的“裸脸”或“裸妆”展品没有参展资格。本次参展作品旨在选拔代表中国盆景的最高水平的展品，所以对参展作品的成熟度、制作技术、艺术水准、配盆、几架、展前美化和修饰等各方面的要求都是历届展览以来最高的。望各省区协会在推荐展品时层层把关筛选，宁缺勿滥。

注：树的高度应从盆面起至顶端绿色（活的部分）位置计算，悬崖盆景飘长应从盆沿算起。

各省市被最后通知入选参展的盆景务必在2013年8月30日前将候选的参展作品树种、拉丁名称、规格、送展者姓名、参展作品照片及拍照时间等信息再次邮寄报送大会组委会秘书处，电子邮箱：2849626593@qq.com。

报名办法：

1. 由协会大展前期考察小组实地考察甄选后上报给“中国盆景国家大展入选评比委员会”。

2. 由本会各地会长、秘书长团队组织当地作品参加大展入选申报工作，各地会员可通过本会各地会长来报名参加大展的入选竞选申报。

3. 当地无会长或无法联系到当地会长的会员也可自行申请参加竞选。具体方式为：发送盆景的正反面数码照片各1张、树种、拉丁名、树干最粗处尺寸、高、长、正反面角度、收藏人名字、收藏人联系电话、通信地址、电子邮箱等资料到评委会报名电子信箱：2849626593@qq.com 。

通过以上3种报名渠道竞选参展的结果将由国家大展的入选评比委员会最终确定。

各地参展作品的运送： 所有盆景展品费用和运输均由送展者自负。

关于国家大展的首席大奖——“中国鼎”超级奖杯的说明

中国盆景国家大展最高总冠军奖项的名称为： 中国鼎。该鼎以具有中国传统文化和历史内涵的中国商代鼎造型为模板，通体采用全玉材质精心制作而成。中国鼎今后将作为一个文化符号向全球传播中国盆景的国家文化形象。

中国鼎大奖的保存和传承机制：“中国鼎”超级奖杯每届的获得者姓名和获奖年度将永久性刻在获奖鼎座上，获奖人有保管此鼎一届的权利，在保留前必须签署保管此鼎防止丢失或损害的全额赔偿协议。下届展览颁奖时此玉鼎持有人必须将此鼎转交给下届新的“中国鼎大奖”获得者保存一届。以此类推，总计3次获得中国鼎大奖的送展人将可永久性保留此具备传世价值的玉鼎。

“2013中国盆景年度先生”： 这是中国盆景的年度个人最高荣誉奖。获得上述“中国鼎”的送展人同时将自动当选中国盆景

界本年度最高的个人荣誉——“2013中国盆景年度先生”。

中国鼎奖杯之外的奖金数额: 奖金总额为8万人民币。

举办时间: 国家大展今后将每两年举办一次。下届中国盆景国家大展的举办时间拟定为2015年。

国家大展的参展盆数: 每届100盆。每届将从全国选取100盆代表中国盆景顶级水平的盆景展品参展。展品在选取时将以历届全国盆景展的金奖作品的平均水平线为入选参考线。

入选盆景的地域范围: 中国盆景国家大展今后将是向全球开放的中国盆景国家展，所有落户于中国的顶级盆景，不分国籍、地域，只要符合国家大展的入选和参展标准，均可报名竞选入赛。全球所有盆景爱好者（包括非会员）均可参加中国盆景国家大展的竞选入赛活动。

评比标准: 中国盆景艺术家协会（CPAA）将在此次大展上推出中国盆景独立的评比标准——包括视觉标准、物理标准、美学标准、哲学和价值标准。

中国盆景国家大展奖项的改革与设计: 本着对国家大展专业和高端的定位，中国盆景国家大展计划对过去的大展评奖标准进行如下三项大幅度的改革：一是减少大奖奖项，除中国鼎首席大奖外，每届国家大展将仅有8盆作品获得“中国盆景国家大展奖”，奖金总额将为每盆1万元人民币；二是大幅度提高入围奖的含金量和评奖等级，所有入选此100盆行列的盆景将获得“国家大展入选奖”，今后在评选申报各类国家级职称或称号时，等级将高于本会举办的历届全国展金奖一个等级；三是大幅度地提高大奖的奖金数额。“中国鼎”奖杯的估价为30万元人民币左右，中国鼎奖杯之外还将发放奖金总额为8万人民币的奖金，“中国盆景国家大展奖”的奖金为每盆1万元人民币。“国家大展入选奖”的奖金为每盆1200元人民币。

运输和送展费用: 今后国家大展所有展品不再提供运费补贴，改为以入选奖奖金的形式给予奖励。

评比方式: 将继续采用2012年在中山古镇的中国盆景精品展上的公开打分制，《中国盆景赏石》将继续公布国家大展的所有评分记录。

关于国家大展申办制的说明: 今后的国家大展将采用申办制模式，由申请人每年提前一年向协会提交申办方案。若有多人提出，中国盆景艺术家协会将采用竞标方式决出最后的申办者。申办单位需提前一年向中国盆景艺术家协会提交竞标方案，在竞办流程中胜出的申办单位或城市将最后获得承办权。由政府提出申办的承办政府所在地城市将获选当年的“中国盆景年度城市（城镇）”的城市名片称号。

其他活动的介绍

中国盆景艺术家协会会员展: 与国家大展同时，将举办 200盆经过层层选拔的高水平盆景参展的中国盆景艺术家协会会员展，与国家大展交相辉映，共同展现中国盆景新时代的高端水准。

其中，大奖奖项为1名，金银铜奖70名，所有金银铜奖均相当于全国盆展的相应评奖等级，详情如下：

“2013中国盆景会员展年度大奖”：1名：奖金1万元。

2013中国盆景会员展金奖（等同于全国展金奖）：10名。

2013中国盆景会员展银奖（等同于全国展银奖）：20名。

2013中国盆景会员展铜奖（等同于全国展铜奖）：40名。

所有金银铜获奖者和入选者均发奖金每盆600元。

会员展的作品的选拔和入围，将与国家大展同步进行，入选会员展的将颁发给“中国盆景艺术家协会会员展入选奖”，今后也将作为中国盆景艺术家协会所有考评活动的等级评比条件之一。与国家大展相随的中国盆景艺术家协会会员展不提供任何运费补贴但将以入选奖奖金的形式给予奖励。

2013中国古盆收藏展: 将展出中国盆景艺术家协会会员收藏的全球最高水平的中国古典盆景盆50件左右。

2013中国盆景艺术家协会会员赏石展: 将展出本会会员收藏的世界一流水平的观赏石50~80件。

其他事项

届时大会将邀请著名盆景大师、专家及盆景界知名人士出席开幕式、参观展览及技艺表演和文化交流活动。

所有参展或参观者交通食宿费用自理，联系食宿者请与大会组委会联系(组委会联系方式另行公布)。

国家大展组委会成立前的前期联系方式:

会务: 马嘉辉, 0760-22389523, 13925300210;

布展: 谢克英, 13928281719;

中国盆景艺术家协会秘书处: 010-58690358;

展品申请参展报名电子邮箱: 2849626593@qq.com;

报名截至日期: 2013年8月10日。

中国盆景艺术家协会

2013 年 6 月

China Ding 2013 China(Guzhen) National Penjing Exhibition Announcement

To: All the members of China Penjing Artists Association and all national Penjing fans

To forge the iconic brand of China National Penjing Exhibition in the aim of facing the whole world and showing the world the national image of China Penjing, the first " China (Guzhen) National Penjing Exhibition " in China Penjing history will be held in Guzhen , Zhongshan City, Guangdong Province during the period from September 29 to October 3, 2013. From now on, all the members of China Penjing Artists Association can register at the neighborhood principals of China Penjing Artists Association for the competition (for example: members of Guangdong Penjing Association can register at Guangdong Penjing Association for the exhibition). And, during China National Penjing Exhibition, numerous significant events will also be launched, and the details are as follows:

Introduction to 2013 China(Guzhen) National Penjing Exhibition

The interpretation and positioning of the name China National Penjing Exhibition: " China (Guzhen) National Penjing Exhibition " is a national Penjing exhibition brand, which is launched by China Penjing Artists Association (CPAA) and Zhongshan Municipal People's Government in 2013, with the positioning (exhibition target) of a national exhibition representing the highest professional Penjing level in China hosted by the National First-class Association.

Hosted by: Guangdong Zhongshan Municipal People's Government

China Penjing Artists Association

Organized by: Zhongshan Guzhen People's Government

Agriculture Bureau of Zhongshan

Forestry Bureau of Zhongshan

Ocean and Fisheries Bureau of Zhongshan City

Guangdong Penjing Association

Co-organized by: Zhongshan Nanfang Green Exposition Garden Co., Ltd.

Supported by: Zhongshan Penjing Association, Zhongshan Guzhen Penjing Association, cooperative associations of China Penjing Artists Association in national provinces and cities

Place: Guzhen (City of Light) Convention and Exhibition Centre, Guzhen Town, Zhongshan City, Guangdong

Date: September 29 to October 3, 2013

Agenda:

September 26~27: Exhibition arrangement

September 28: Award appraisal and guest registration

September 29: Opening ceremony and exhibition visit by responsible units, members, and guests

6:30 PM of September 29: Annual Night of China Penjing

September 30: Open to the public; guests' visiting exhibition and demonstration show in the morning and academic activities in the afternoon.

October 1~3: Exhibition show

October 4: Show over

Exhibit specifications and selection

The standard for specifications and sizes of exhibits in China National Penjing Exhibition are: Within 1.5m high (the literati tree can be within 1.6m high, and its appraisal standard can be determined by votes of the judge group); the pot within 1.8m long, and the cliff branch within 1.5m long.

All exhibits shall be supplied with the corresponding shelf (generally 10~20cm), and the Penjing's pre-exhibition trimming, beautification, pot preparation, and shelf collocation shall be listed in its marking. The "naked-face" or "naked-look" exhibits without elaborate pre-exhibition trimming are unqualified for the exhibition, and the Exhibition is intended to select the exhibit representing the highest Penjing level in China; on this account, the requirements on the exhibit's maturity, making technology, artistic standard, pot preparation, shelf, and pre-exhibition beautification and trimming are the strictest. It is hoped that, all province-level and district-level associations can make checks at all levels when selecting the exhibits.

Note: The tree height shall be calculated from the pot surface to the green position (live part), and cliff branch length shall be calculated from the pot edge.

All provinces and cities are required to mail the information including the species, Latin name, specifications, participant's name, exhibit photo and photographing time of the candidate exhibit to Secretariat of the organizing committee(E-mail: 2849626593@qq.com) before August 30, 2013.

How to Apply

1. After the preliminary investigation group of the Exhibition makes the site investigation, it will report the result to the "appraisal committee of China National Penjing Exhibition".

2. The presidents of all district and region Associations and secretariat group will organize the application of the exhibits, and the members can register at the corresponding presidents for the exhibition.

3. If there is no president locally or it is unable to contact the president, the member can apply for the exhibition. The specific method: Send the information including 2 digital photos for each of the Penjing's front and back sides; species, Latin name, dimensions of the thickest trunk, height, length, angle of the front and back sides, participant's name, telephone number, address, and E-mail to the appraisal committee (E-mail: 2849626593@qq.com).

The exhibit results of the above 3 registration methods will be finally determined by the appraisal committee of China National Penjing Exhibition .

Transportation of all exhibits: All expenses and transportation of the Penjing will be undertaken by the participant.

About the Top Award of China National Penjing Exhibition —Interpretations of the Super Cup of "China Ding"

The name of the top champion award of China National Penjing Exhibition : China Ding. China Ding takes the Shang Dynasty Ding with Chinese traditional culture and history as the sample, which is elaborated with the jade wholly. In the future, China Ding will be made as a culture symbol to globally spread the national culture image of China Penjing.

Preservation and inheritance mechanism of China Ding Award: The names and award years of all winners of the Super Cup of "China Ding" will be permanently marked on the base of the "Ding"; the winner is entitled to keep this Ding for one term; however, he shall sign the full compensation agreement regarding losses or damages. Upon the next Exhibition, the Ding holder shall deliver the Ding to the next winner for 1-term keeping. By parity of reasoning, if the winner wins the Ding for three times, he can permanently keep it.

"2013 Annual Man of China Penjing": It is the highest annual individual award of China Penjing. The participant winning the above "China Ding" will be automatically selected as the highest annual individual award of China Penjing—"2013 Annual Man of China Penjing".

In addition to the China Ding Cup, the winner can get the bonus with an amount of RMB 80,000.

Exhibition hosting time: China National Penjing Exhibition will be hosted once every two years in the future. It is planned to hold the next China National Penjing Exhibition in 2015.

Quantity of exhibits for China National Penjing Exhibition : 100 exhibits for every Exhibition. In the Exhibition, 100 exhibits will be selected from all exhibits for competition. Upon selection, the average level of the works winning gold prize of the previous Exhibitions will be made as the standard.

Territorial scope of the selected Penjing: In the future China National Penjing Exhibition will be the one that opens to the world, and all top-level China Penjing up to the selection and exhibition standard of China National Penjing Exhibition can register for the exhibition regardless of their nationalities and regions. All Penjing fans at home and abroad (including the nonmembers) are entitled to participate in competition activity of China National Penjing Exhibition .

Appraisal standard: China Penjing Artists Association (CPAA) will introduce the independent appraisal standards for China Penjing, including the visual standard, physical standard, aesthetic standard, and philosophy and value standard.

Reform and design of the award of China National Penjing Exhibition:

For a professional and high-end positioning of China National Penjing Exhibition, China National Penjing Exhibition plans to make following significant reforms comparing with the previous appraisal standards: The first is to reduce the award items; except the top award of China Ding, only 8 exhibits will win the "Award of China National Penjing Exhibition ", with the bonus of $1667 for each exhibit; the second is to greatly improve the gold content and appraisal level of the nominee awards, all selected 100 exhibits will win the "Nominee Award of China National Penjing Exhibition"; when applying for the national title, its grade will be one grade higher than the gold award of the previous Conventions; the third is to largely increase the award bonus: The Cup of "China Ding" is about $50000; in addition to the Cup, the bonus with an amount of $13000 will be distributed. For the "Award of China National Penjing Exhibition ", the bonus will be $1667 for each exhibit; for the "Nominee Award of China National Penjing Exhibition", the bonus will be $200 for each exhibit.

Transportation and participation expenses: In the future, no transportation subsidies will be given to the exhibits for China National Penjing Exhibition, which will be changed to the bonus for the "Nominee Award of China National Penjing Exhibition".

Appraisal method: The public marking system applied for 2012 China Penjing Exhibition in Guzhen Town, Zhongshan will be applied, and all marking records of China National Penjing Exhibition will be published in the China Penjing & Scholar Rocks.

Interpretation of the application system of China National Penjing Exhibition: Future China National Penjing Exhibition will be applied with the application system, and the applicant shall submit the application scheme to the Association one year in advance. If applied by multi-persons, the China Penjing Artists Association will use the competitive bidding method for final determination. The applicant shall submit the competitive bidding scheme to China Penjing Artists Association one year in advance, and the final winner will get the final hosting right. The city where the hosting government is located will get the title of the "Annual City (Town) of China Penjing".

Introduction to Other Activities

China Penjing Member Exhibition of Chinese Penjing Artists Association: In addition to China National Penjing Exhibition, China Penjing Artists Association Member Exhibition with 200 high-quality Penjing exhibits will be held, and both them can reflect the high level of new China Penjing times.

There are 1 annual award and 70 gold and silver awards, and the gold and silver awards equal to the corresponding awards of the national Penjing Exhibition . The details are as follows:

"Annual Award of 2013 China Penjing Member Exhibition":1, with the bonus of $1667.

Gold award of 2013 China Penjing Member Exhibition (equal to the gold award of national Penjing Exhibition): 10.

Silver award of 2013 China Penjing Member Exhibition (equal to the silver award of national Penjing Exhibition): 20.

Bronze award of 2013 China Penjing Member